# Internet of Energy for Smart Cities

# Internet of Energy for Smart Cities

## Machine Learning Models and Techniques

Edited by

Anish Jindal, Neeraj Kumar and
Gagangeet Singh Aujla

CRC Press
Taylor & Francis Group
Boca Raton London New York

CRC Press is an imprint of the
Taylor & Francis Group, an **informa** business

First edition published 2022
by CRC Press
6000 Broken Sound Parkway NW, Suite 300, Boca Raton, FL 33487-2742

and by CRC Press
2 Park Square, Milton Park, Abingdon, Oxon, OX14 4RN

Library of Congress Cataloging-in-Publication Data

ISBN: 978-0-367-49775-0 (hbk)
ISBN: 978-0-367-49955-6 (pbk)
ISBN: 978-1-003-04731-5 (ebk)

Typeset in Nimbus
by KnowledgeWorks Global Ltd.

# Contents

## SECTION I  Overview

# SECTION II   Smart Grids

# SECTION III   Internet of Energy (IoE)

*Ash Mohammad Abbas*

*Manju Lata and Vikas Kumar*

*Rania Salih Abdalla, Sara A. Mahbub, Rania A. Mokhtar,
Elmustafa Sayed Ali, and Rashid A. Saeed*

# SECTION IV   Machine Learning Models

*Dristi Datta and Nurul I. Sarkar*

## SECTION V    Case Studies and Future Directions

**Chapter 10**  Intelligent Control System for Smart Environment Using Internet of Things ...................................................................... 265

*Chintan Bhatt, Riya Patel, Siddharth Patel, Hussain Sadikot, Akrit Khanna, and Esha Shah*

**Chapter 11**  Pathway and Future of IoE in Smart Cities ............................. 277

*Sharda Tripathi and Swades De*

# Preface

With the advent of new technological breakthroughs, many concepts and techniques have evolved over the years. These advancements have revolutionized many application areas; one of which has led to the smart city era. The key aspect of the smart city is to integrate information and communication technologies (ICT) for providing various services, such as smart healthcare, smart connectivity, smart transportation, smart resource management, smart governance, and public safety. Out of the aforementioned services in a smart city, energy management, which primarily focuses on utilization of the energy resources efficiently, is a paramount concern. Energy is one of the most valuable resources of the modern era which needs to be utilized in such a manner so that there exists a balance between supply and demand. Even with the growing interest toward smart grids, the power industry has always faced a challenge of amalgamating the data and underlying systems running various softwares and complex protocols to realize the desired performance and expected capabilities. However, the vision of an intelligent, cloud-based distributed energy system and its management have the potential to answer most of these challenges which arise from time to time. Though it has not been long since Internet of Things (IoT) technologies entered the energy sector, it cannot be denied that IoT is now everywhere, so the utmost need is to harness the possibilities that these technologies can create or bring. In today's industry, the so-called "Smart Grid" has to look further into the future with digitization as a key enabler for the energy industry or the connected energy world. The functionality offered by cloud-based computational platforms for processing data and running co-developed grid applications enables the power grid to operate as a single energy ecosystem. In this connected energy ecosystem, the data are analyzed to optimize the grid and help to control and operate the resources in a more effective, sustainable, and resilient manner through the usage of automation and digitization. This networked energy ecosystem, or popularly known nowadays as the Internet-of-Energy (IoE), is a web of different energy infrastructures operating across various levels (such as generation units, smart meters, storage, distribution units, etc.). The individual grid devices and infrastructure communicate and coordinate with each other to collect data, process (organize or analyze) it, and thereafter utilize the extracted information by distributing it across the network quickly.

IoE enables (and empowers) the consumers as well as prosumers to connect and coordinate the supply and demand among themselves in an autonomous manner using machine learning-based models such as managing energy demand by using forecasting methods to predict future energy supply and demand patterns. Let us consider an example of meter data management to understand the role and applicability of IoE. The cloud-based application automates meter data management in line with business processes to generate actionable information using the digital world. This allows power utility to address the longstanding demand (like meter-to-cash billing) along with the provision to leverage the interval or time-of-use data to support the mission critical applications. Using IoE, utilities can manage and operate all smart energy infrastructure processes and data acquisition systems, communicate with all grid devices, receive error/problem notifications, and manage customer side metering points (like electric vehicles or solar panels). Talking about the substation (or a more abstract) level, IoE can help automate the asset inventory lists, compare the actual and targeted values across various versions and settings, and provide remote support and services to the operators, such as supporting security patch management. It also helps to differentiate the data streams related to the control functions and asset management at the substation level. The asset information can be analyzed and analytical models can be executed in real time to ensure a proactive intervention or substitution. Moreover, the operators can utilize the data to provide real-time forecasting of a renewable energy generation and then integrate renewable generators in a cost-effective manner without the need for grid extension. With connected energy, they can also plan efficiently for energy market resources, improve profits through energy trading, and handle power outages in an intelligent manner.

However (as in most of the cases), the benefits are always accompanied by the challenges. This is true for IoE as well since it has some technical constraints. The most prominent challenges include the secure and reliable transmission and efficient processing of data as it is generated in huge volume during different grid operations. In addition, the IoE ecosystem involves the integration of a multitude of grid-cloud devices developed (and deployed) by many vendors/manufacturers, so interoperability and scalability are yet another issue which pose a stiff challenge for the true realization of IoE. In the era of Industry 4.0, we are steadily moving toward a future where the contemporary grid would be driven through the combination of machine learning models executing intelligent devices through a high-speed communication network providing a near to real-time processing.

In this regard, machine learning approaches have the capabilities to learn and adapt to the constantly evolving demands of this large IoE network. These approaches have already proved their worth in other technical fields such as image processing, deep learning, pattern recognition, etc. The self-learning features in machine learning approaches make them the perfect candidate solutions for catering to the emerging needs of smart cities. Therefore, the focus of this book is on using the machine learning approaches to present various solutions for IoE networks in smart cities. This book also provides in-depth knowledge to build technical understanding for the reader to pursue various research problems in the field of machine learning in smart cities for solving energy-related problems. The readers will also benefit from detailed explanation of used concepts in order to build the machine learning models. Moreover, case studies in smart cities and their solutions are provided to be more correlated to real-life scenarios.

Keeping in mind the above discussion, this book is divided into five sections. The first section provides the background information and builds up the research problems that persist in smart cities' energy ecosystems. The second section focuses on the Smart Grid and related technologies, including the transition from a conventional grid to a modern connected grid. The third section discusses the IoE perspective, associated technologies, design challenges, underlying architecture, future challenges, and IoE applications. The fourth section includes technical details of various machine learning solutions that can be employed in order to solve the problems and challenges related to the IoE ecosystem in smart cities. The last section provided case studies to solve various problems (including energy management, intelligent control systems and load management) related to IoE using machine learning models. The book ends with a detailed discussion on the future pathways and directions of IoE in smart cities.

**Anish Jindal, Neeraj Kumar, Gagangeet Singh Aujla**
Editors

# Editors

**Anish Jindal**
University of Essex
United Kingdom

**Neeraj Kumar**
Thapar Institute of Engineering and Technology
India

**Gagangeet Singh Aujla**
Durham University
United Kingdom

# Contributors

**Ash Mohammad Abbas**
Aligarh Muslim University
India

**Rania Salih Abdalla**
Sudan University of Science and
    Technology
Sudan

**Elmustafa Sayed Ali**
Sudan University of Science and
    Technology
Sudan
Red Sea University
Sudan

**Chintan Bhatt**
Charotar University of Science and
    Technology
India

**Haba Cristian-Gyozo**
Faculty of Electrical Engineering Iasi
Romania

**Dristi Datta**
Varendra University
Bangladesh

**Swades De**
Indian Institute of Technology Delhi
India

**Federica Foiadelli**
Politecnico Di Milano
Italy

**B. Isaías Lima Fuly**
Federal University of Itajubá
Brazil

**Akrit Khanna**
Charotar University of Science and
    Technology
India

**Vikas Kumar**
Chaudhary Bansi Lal University
India

**Manju Lata**
Chaudhary Bansi Lal University
India

**Michela Longo**
Politecnico Di Milano
Italy

**Sara A. Mahbub**
Sudan University of Science and
    Technology
Sudan

**Seyed Mahdi Miraftabzadeh**
Politecnico Di Milano
Italy

**Rania A. Mokhtar**
Taif University
KSA
Sudan University of Science and
    Technology
Sudan

**B.K. Panigrahi**
Indian Institute of Technology Delhi
India

**Riya Patel**
Charotar University of Science and
    Technology
India

**Siddharth Patel**
Charotar University of Science and
    Technology
India

**Paulo F. Ribeiro**
Federal University of Itajubá
Brazil

**Hussain Sadikot**
Charotar University of Science and
    Technology
India

**Rashid A. Saeed**
Taif University
KSA
Sudan University of Science and
    Technology
Sudan

**Rafael S. Salles**
Federal University of Itajubá
Brazil

**Nurul I. Sarkar**
Auckland University of Technology
New Zealand

**Esha Shah**
Charotar University of Science and
    Technology
India

**Sumedha Sharma**
Indian Institute of Technology Delhi
India

**Sharda Tripathi**
Politecnico di Torino
Italy

**Ashu Verma**
Indian Institute of Technology Delhi
India

# Section I

## Overview

# 1 Smart City: The Verticals of Energy Demand and Challenges

*Sumedha Sharma*
Centre for Energy Studies, Indian Institute of Technology Delhi

*Ashu Verma*
Centre for Energy Studies, Indian Institute of Technology Delhi

*B.K. Panigrahi*
Department of Electrical Engineering, Indian Institute of
Technology Delhi

## CONTENTS

## 1.1  INTRODUCTION

Energy forms the basis for existence and sustenance of life. Interactions and dynamics of the ecosystem are attributed to a complex interplay of different forms of energy. Our dependence on energy is such that the growth and development of nations is evaluated based upon energy-derived indicators. These energy indicators enable policy-makers to assess growth and evolution and introduce reforms and directives to regulate the global energy sector. Among the major energy carriers, electricity is one of the most commonly and widely used forms of energy which is directly utilized by the end-users, including residential, commercial and industrial consumers. Over the years, the electrical power system has gradually and drastically evolved in terms of its structure, operation and control practices. The quantum of electrical energy in circulation in the power system has increased owing to technological developments, growing urbanization and modernization. It is estimated that every day humans consume more than a million *terrajoules* of energy, and this demand is expected to rise by 48% from 2012 to 2040 [9].

Consistent efforts to ensure energy security and energy access require power generation to increase at the same rate of increase in demand. However, at present most of this energy is produced using fossil fuels in large scales. Rising and prolonged use of fossil fuels for global energy generation has emerged as the single largest contributor to greenhouse gases. As a consequence, the world's five hottest years on record have occurred since 2015, thus, raising threats of climate change. The eventual threats to global weather patterns and awareness regarding high carbon footprints, along with the depleting levels of fossil fuels have led to the integration of renewable energy sources (RES) in the existing energy system. RES, in contrast to conventional sources of energy, are carbon-free energy sources which replenish themselves within the human time-frame, such as solar, wind, and biomass. Thus, RES integration has been widely identified as the way forward for the global energy scenario.

In view of impending global issues, the United Nations (UN) designed the millennium development goals which are a set of eight time-bound international development goals established in the year 2000 [37]. These focused upon holistic global development with emphasis on pertinent issues such as poverty eradication, gender empowerment, health and education for all, with a 15-year deadline to achieve the set goals. While the millennium development goals proved to be successful and instrumental in combating several prominent issue, the need for effective handling of global warming and climate change was increasingly and unanimously realized worldwide. The experience and success of millennium development goals paved way for the adoption of a sustainable approach to global development, which led to the foundation of the sustainable development goals (SDGs) [35]. Thus, in the year 2015, the millennium development goals were superseded by the SDGs, which were designed as a "blueprint to achieve a better and more sustainable future for all". Accordingly, 17 SDGs were set by the United Nations General Assembly in 2015 at the COP21 Paris Climate Conference [33]. The essence of the SDGs is the incorporation of efforts to reduce global carbon emissions and tackle the impending risks of climate change. The SDGs have been designed with an integrated approach which recognizes that action in one area will affect outcomes in others, and that development must balance social, economic and environmental sustainability.

Driven by the sustainable development goals (SDG), *smart cities* have been conceptualized which aim at the provision of urban ecosystems with adequate infrastructure for holistic development, and enhancing the quality of life of its individuals. To meet the increasing energy needs, nations are gradually moving toward adoption of cleaner energy sources with lower emissions to minimize carbon footprints, while improving the physical quality of life. Since smart cities focus on quantitative and qualitative metrics of development and quality of life, the concept is relative to the state of development of a geographical area and may vary as countries develop with time, and also may differ from country to country, or even within one country, from state to state, and city to city. The principle guidelines are, however, governed by the directives of *SDG 11*. What appears common to most of these definitions is the recognition of the need to consider an innovative and forward-looking use of the infrastructural systems, with positive outcomes on climate, environment, and social and economic conditions.

## United Nations Sustainable Development Goal 11: Sustainable Cities and Communities

SDG 11 is focused upon making cities *inclusive, safe, resilient and sustainable*. It was estimated that in 2018 nearly 4.2 billion people, which account for 55% of the global population lived in cities. This number is expected to grow to 5 billion by 2030 owing to rapid urbanization and migration. This highlights the need to equip urban settlements with solutions to combat the major issues that humanity is faced with. Among the pressing challenges faced by cities are the high levels of energy consumption. It has been estimated that cities occupy around 3% of the Earth's land, however, their energy consumption accounts for 60-80% of the total energy consumption, and contribute to 75% of the global carbon emissions. The United Nations has thus, identified that smart cities hold the key to sustainable development. Accordingly, smart cities and communities have been included among the 17 global goals set by the United Nations in 2015, which are intended to be achieved by 2030 [2, 34].

Energy security and access are key aspects of a smart city. Globally 860 million people do not have access to electricity. Ensuring adequate supply of power to the global population would require increasing the generation at the same rate as demand, and suitable transmission resources. It is noteworthy that the existing electricity transmission infrastructure that transfers bulk power is gradually ageing and losing its reliability, and establishing a new transmission system is highly capital intensive. While meeting rising loads by increasing generation through fossil-fuel-based central power stations is constrained by high $CO_2$ emissions and limited availability of fossil fuels, RES integration is faced with grid-integration challenges owing to specific peculiarities of these systems. RES are inherently intermittent and variable in nature. Interconnecting these sources to the main grid leads to issues of power quality, frequency and voltage regulation. However, being distributed in nature, energy from RES can be harnessed locally closer to the utilization end, thereby reducing transmission losses. Furthermore, utilization closer to the consumption point minimizes costs associated with power transmission and distribution. Additionally, interconnection issues at the distribution level are fairly small scale in comparison to grid-level issues, and can be mitigated with suitable distribution-level corrective techniques.

Accordingly, continuous efforts are being made in smart cities for development of methods and technologies for harnessing energy through cleaner and sustainable sources [6, 23]. This has led to the emergence of *smart energy systems* which focus on integration of multiple energy-producing and energy-consuming sectors with an objective to identify and develop coherent sustainable and optimal solutions for future energy requirements. Incorporation of information and communication technology in smart energy systems enables suitable control of multiple generation, energy storage and load resources through generator control actions and demand-side control measures such as demand response and demand side management. Thus, smart energy systems essentially focus on identifying sub-sectoral synergies among electrical, thermal and gas systems, with an attempt to derive low-carbon and sustainable infrastructural designs, and operational strategies. Some prominent sub-sectoral synergies include the following:

1. District heating and cooling offers significant energy savings which promotes development of electricity and heat-based technologies for space heating/cooling in buildings and microgrids.
2. Waste heat emitted from industries and power plants can be used for space heating/cooling through district thermal distribution networks and heat pumps.
3. Combined production of electricity and heat through combined heat and power (CHP) units improve utilization efficiencies of the generating system, which is otherwise very low since a significant amount of energy is lost in multiple conversion processes.
4. Thermal energy storage, being cheaper and more efficient in comparison to electrical storage, enables flexible operation of space heating/cooling in buildings and microgrids.
5. Thermal loads served by electricity may assist the electrical grid as balancing resources through demand response.
6. Biofuel and biogas are clean fuels for CHP generation, thus, emerging as strategic energy players in rural areas where availability of biomass is abundant.
7. Electric vehicles have emerged as clean alternatives to fuel in the transportation sector, while also providing grid-balancing services.

In accordance to the form of primary energy flow in a network, the energy sector is broadly characterized into electricity, thermal and gas grids, discussed as follows.

1. **Smart electricity grids**: Primary form of energy in these networks is electricity. This is generated by conventional or renewable generation sources, transmitted and distributed over an electrical network, and consumed by loads or stored in batteries, at utilization points. Communication and control infrastructure facilitate the integration of flexible loads, which are largely constituted by thermal loads such as air conditioners and heat pumps, and electric vehicles, as well as intermittent and variable renewable energy sources.

2. **Smart thermal grid**: The primary form of energy flow in such networks is heat (thermal energy). Through suitable distribution and conversion infrastructure, the thermal energy may directly be used for district heating and cooling needs. This also facilitates the utilization of thermal energy storage systems, which act as diurnal buffers for thermal energy; or seasonal thermal energy storage, which are long-term thermal storage systems, wherein heat is stored at depths in the earth's layers in one season, to be used during the other season.

3. **Smart gas grids**: The gas network connects the electrical, thermal and transportation sectors, thus emerging as enablers of complete energy systems. Gas distribution networks and storage systems impart additional flexibility to power generation through dual-fuel engines, gas turbines, dual-fuel appliances such as heat pumps. Further, gas majorly sustains the transportation sector in gaseous as well as liquid fuel forms.

Accordingly, electrical, thermal and gas networks are integrated in an approach which is commonly referred to as multi-energy system or integrated energy system. The prime focus of designing such systems is to achieve most economic, reliable and secure operation of the individual energy systems as well as the integrated energy system. This has led to design and development of operational strategies and system optimizers, which identify the inter-sectoral synergies and coordinate the multiple system components through incorporation of suitable communication and control infrastructure.

---

### Smart Energy Systems

"A smart energy system is defined as an approach in which smart electricity, thermal and gas grids are combined with storage technologies and coordinated to identify synergies between them in order to achieve an optimal solution for each individual sector as well as for the overall energy system" [27].

---

It is noteworthy that distributed energy system components have shorter response times and lower inertia and are their control is, thus, critical in assisting small-timescale/online grid operations. The penetration of RES and storage sources in distributed systems results in bidirectional flow of power which enhances distribution system complexity and necessitates the development of analytical tools to assess real-time performance. Owing to the prominence of spatial and temporal variations in distributed energy systems, multiple parallel computational processes are required to enable efficient control. Thus, their operation arises various challenges which include coordinated control of the spatially distributed processes, real-time decision making and control, dynamics of demand response programs, mitigating temporal variations in states and forecasts and big data management. This necessitates the development of suitable control algorithms which are capable of performing multi-timescale coordination owing to different response and control times of each component.

Moreover, real-time operation is associated with high levels of uncertainties due to intermittent renewable energy sources and forecast errors [4]. Thus, real-time grid analytics require uncertainty handling with appropriate data processing, secure data handling and control with short lead-times. The backbone for operation of such a system is a well-developed architecture and communication system that enables fast data transfer to facilitate real-time computation and control. Thus, with the evolution of the smart energy systems, data storage, transfer and its handling has emerged as a big issue. In view of the aforementioned discussion, this chapter aims to provide a comprehensive view to the readers about demand response, its electrical and computational aspects, along with the development of online algorithms for smart distribution systems, detailed operational aspects, and associated challenges.

## 1.2 SMART ENERGY DISTRIBUTION

Among the various forms of energy flowing in a network, electricity has been identified to play a prominent role owing to its central role in modern life, and increasing contribution in the total energy use, popularly known as electrification of the society. Consequently, the electric grid is undergoing a shift in generation mix with increased integration of RES in the electrical power system. Due to the continuously increasing loads, the electric grid is constantly stressed while trying to meet the generation-load balance at all times. The electrical power transmission infrastructure, in its present form, in most parts of the world, has been in existence for decades. Consequently, a significant portion of the transmission infrastructure is faced with the challenge of accelerated ageing due to higher loads for increased duration. This poses greater risks of equipment failures, and their cascaded effects may eventually lead to brownouts and blackouts. While this raises concerns over energy security and access, augmenting the transmission infrastructure is a highly capital-intensive task.

Rising average and elevated peaks in a load-duration curve correspond to adverse effects on the power system. These include higher losses, congestion, capacity challenges and operating limit violations. Network losses are inevitable and are governed by the *Joule* effect, represented by *ohmic* $(I^2R)$ loss. It is, thus, evident that ohmic losses are higher in distribution systems which have higher $R/X$ ratio, as compared to transmission networks which have lower values of the $R/X$ ratio. It is estimated that distribution system line losses constitute 13% of total power generation [40]. This highlights the fact that management of power flow in a distribution network is imminent for achieving favorable network performance, with lower losses, higher reliability of power and improved energy security without having the need for huge capital investments for transmission network augmentation.

Distribution system management has led to the emergence of the concept of *smaller energy systems* at different scales of sizes. These include local power distribution networks, in contrast to the large interconnected power system, which depending upon the system capacity may be categorized as minigrids, microgrids, nanogrids, picogrids, etc. [15]. Understandably, the categorization and sizing of smaller energy systems differ from country to country as there exists no standard uniformly accepted categorization. For instance, in India a minigrid is sized 10-200 kW, while a microgrid is smaller than 10 kW, and nanogrids and picogrids are even smaller networks. However, in practice, the term microgrid is generally used to refer all smaller energy systems despite the size categorization. There exists a general consensus over a few common definitions of microgrids which have been discussed below.

---

## Microgrid Definitions

*CIGRE C6.22 Working Group:* "A microgrid is a group of interconnected loads and distributed energy resources within clearly defined electrical boundaries that acts as a single controllable entity with respect to the grid. A microgrid can connect and disconnect from the grid to enable it to operate in both grid-connected or island-mode" [12].

*U.S. Department of Energy Microgrid Exchange Group:* "A microgrid is a group of interconnected loads and distributed energy resources within clearly defined electrical boundaries that acts as a single controllable entity with respect to the grid. A microgrid can connect and disconnect from the grid to enable it to operate in both grid-connected or island-mode"[10].

*World Bank:* "Isolated, small-scale distribution networks typically operating below 11 kilovolts (kV) that provide power to a localized group of customers and produce electricity from small generators, potentially coupled with energy storage system" [32].

*Lawrence Berkeley National Laboratory:* "A small electrical domain connected to the grid of no greater than 100 kW and limited to a single building structure or primary load or a network of off-grid loads not exceeding 5 kW, both categories representing devices (such as DG, batteries, EVs [electric vehicles], and smart loads) capable of islanding and/or energy self-sufficiency through some level of intelligent DER management or controls".

---

In common practice, microgrids or other smaller energy systems consist of multiple distributed generation sources such as solar photovoltaic (PV) systems, wind turbine generators, fuel cells and diesel generators. These are also commonly referred to as distributed energy resources (DER). Additionally, microgrids may also consist of electrical energy storage systems such as battery banks and supercapacitors, which act as energy buffers and perform the task of energy shifting over periods of time and mitigating RES uncertainty and intermittency. Electrical loads, owing to their flexible operation, are increasingly being considered as *demand resources*. Smart microgrids are integrated with an additional Information and Communication Technology layer, which enables bidirectional communication among the system components. The presence of multiple devices operating in different timescales and having different response times envisages the need for suitable control systems. Accordingly, these systems are designed with computational and control capabilities, which impart the characteristics of intelligence, proactiveness and autonomous control. The computational element is essentially an optimizer, which may be referred to as system operator or energy manager. It receives data from sensors and computes an optimal decision which is then communicated to the actuators for practical implementation.

## 1.2.1   DEMAND RESPONSE

The fundamental goal of every power system is to continuously minimize the generation-load imbalance, in order to meet the system loads at an acceptable reliability level while ensuring system security and several operational constraints of the various power system components [3, 5]. Traditionally, the operation of large interconnected power systems was driven by the need to satisfy loads, commonly referred to as "load-following" mode of operation. The generation and load control decisions in any operating power system is obtained as an instruction by a system operator, at different timescales of operation. However, load control decision by the system operator traditionally pertained to complete shut-down of aggregated load at the substation-level. This type of load control, however, is forced upon and does not impart an active role to the users.

With the development of advanced controllability at the appliance level and adequate communication infrastructure, the demand-side has been identified as a potential active resource for the power system. Increased penetration of DER, controllable loads and energy storage systems enable the customers to respond to grid requirements by optimally managing the power flow at the distribution end [17]. Accordingly, the operational philosophy of the power system has gradually modified from being "generation follows load" to "load follows supply". This philosophy has been facilitated by the emergence of smart energy systems wherein the *plug-and-play* feature and *load responsiveness* are important system characteristics. This has led to the appropriate coining of the terms *demand response* and *demand side management*, which have emerged as the focus of power system planners and operators throughout the world.

The Northwest Power and Conservation Council defines demand response (DR) as "*a non-persistent intentional change in net electricity usage by end-use customers from normal consumptive patterns in response to a request on behalf of, or by, a power and/or distribution/transmission system operator. This change is driven by an agreement, potentially financial, or tariff between two or more participating parties*". In simpler words, demand response refers to strategic actions by distribution companies to motivate users to reduce or shift their energy consumption from peak to off-peak hours. It is generally a pre-arranged arrangement between the distribution company or system operator and load-aggregators or consumers. It provides users with an opportunity to manage their power consumption and reshape their load profiles in response to varying electricity prices or incentives offered by the operator.

Around the world, several independent system operators (ISO) and utilities are operating DR programs, which include California ISO, New England ISO, Pennsylvania-New Jersey-Maryland and Pacific Gas & Electric and Southern California Edison. The two major types of DR programs in practice are *price-based* and *incentive-based* DR programs, which have been discussed below [43].

1. Price-based DR: also known as market-led (economic) DR, refers to programs in which users vary their loads in response to fluctuating electricity prices. The various time-varying prices which may be offered are defined as follows:
   a. Time-of-use (TOU) price: is the electricity pricing mechanism which offers different unit prices at different time blocks. Common TOU rates are peak/off-peak electricity prices which vary according to time of the day, or seasonal price variations.
   b. Real-time price (RTP): refers to a dynamic pricing mechanism wherein the electricity prices are updated at smaller time intervals such as 1 hour or 15 minutes. This reflects marginal variations in the cost of electricity production during the corresponding time interval.
   c. Critical peak price (CPP): is a special pricing mechanism that is designed with an objective to balance demand and supply on certain critical load days. Thus, during normal periods while TOU is in effect, during certain constrained days (critical peak days) the user may either reduce energy consumption or pay the critical peak day rates.
2. Incentive-based DR: also known as system-led (emergency/reliability), DR refers to programs wherein customers reduce their demand during emergency events, such as when the system stability is jeopardized or during the period of price spikes.
   a. Direct load control (DLC): refers to DR programs in which the operator remotely controls user loads and performs direct and complete shut down of the loads.
   b. Interruptible/curtailable DR: refers to the type of DR programs where users volunteer for load reduction during system contingencies in exchange for discounted electricity price or bill credit.
   c. Demand bidding/buyback (DB): refers to the DR mechanism that imparts an active role to users in the competitive electricity market by offering opportunities to obtain financial incentives for modifying electricity consumption though bidding.

## 1.2.2   DEMAND SIDE MANAGEMENT

While DR refers to short-term variations in electricity consumption which are induced by the system operator, long-term interventions to achieve reduction in energy demand are referred to as demand side management (DSM) techniques. Essentially, DSM is the co-ordination of loads on the demand side, and set of activities that helps utility and its customers to modify the shape, and reduce peak to flatten the load curve [21, 24]. It is used to control the way consumers use electricity to achieve energy savings with higher efficiency in energy use. By the use of DSM, improved transmission and distribution grid efficiency can be achieved, while promoting effective utilization of RES, and refine stability and efficiency of the electric grid. The field of DSM encompasses varied techniques and algorithms. An illustration of load shapes with application of DSM techniques is given in Figure 1.1. There are six main DSM techniques, namely, load shifting, peak clipping, conservation, load building, valley filling, and flexible load shape [25, 26].

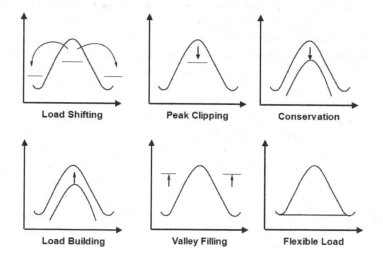

**Figure 1.1**   Various DSM techniques

Load shifting technique shifts deferrable loads from peak hours to off-peak hours resulting in lower peak and flatter curve. In peak clipping, load from peak hours is reduced by encouraging consumers to decrease their power consumption. It is a direct load control technique. Conservation refers to continuous reduction in power consumption by users. Consequently, reduction in complete load curve is achieved by motivating consumers to use more energy efficient appliances. Valley filling includes increase in loads during off-peak hours to a achieve flatter load curve. It is done by increase in energy consumption. This technique may be required when the average price is lower than the cost of load building in the off-peak hours. Load building technique is used in case of increased demand in electricity to optimize daily responses beyond valley filling. It is required in case of surplus energy production.

Flexible load technique focuses on smart grid reliability, allowing consumers to use some power at lower than normal rates. The consumers which agree to use flexible loads during critical periods are facilitated with various incentives.

DR and DSM measures, as discussed above, may collectively be referred to as intelligent load management or load control techniques. Over the years, these have proven to be important tools in the operation of modern power systems. With the emergence of smart energy systems, these intelligent load management techniques have assisted in reliable and stable system operation by mitigating uncertainties which are inherent in high RES penetrated systems, improving load factor, imparting an active role to consumers by enabling load reduction for peer-to-peer transactions, facilitating energy trading within the system, as well as exporting power to the utility, reducing the overall energy consumption, providing ancillary services to the distribution networks, among other operations of the energy systems [17]. A detailed discussion of energy system operations in different timescales is presented in the following section.

## 1.3 REAL-TIME GRID ANALYTICS AND DATA MANAGEMENT

The modern energy systems are an amalgamation of multiple intermittent RES, smart devices and controls and storage technologies, interfaced with information and communication infrastructure. Owing to the extensive structural and operational modifications, the "smarter" energy systems are faced with challenges of bidirectional flow of power at multiple timescales. Coordinating the bidirectional power flow with temporally distributed power system operations, as well as spatially distributed generation of resources is a challenging task. Furthermore, the inherently intermittent and uncertain RES create real-time challenges owing to unforeseen variations in generation forecasts. Additionally, the smart energy system operations are associated with huge amount of data exchange among the various system participants. Large volumes of data, flowing at a high rate creates challenges of efficient data handling and storage. Accordingly, this section discusses in detail the real-time dynamics of energy system operations, high-volume data handling, and effective uncertainty mitigation through suitable real-time algorithms capable of small-lead time operation.

### 1.3.1 ENERGY SYSTEM OPERATIONS

Any power system, including the large interconnected systems or smaller energy systems such as microgrids, includes several layers over which the system is planned and operated in different timescales. An illustration representing the various layers of energy system operations is shown in Figure 1.2. As can be seen from the figure, the first layer refers to the *planning* process which identifies plans for integration of new system components such as capacitor banks and feeders, in view of the rising load and generation over a forecast duration, which might be a few months to several years. This is followed by *scheduling* which refers to short-term decision making such as unit commitment, economic dispatch and time-ahead DR. This process is usually dependent upon economic objectives, while several other types of network-related objectives may also be used. The *optimal schedule* is computed over a period of few days to few hours/minutes, until *gate closure* of the utility. Depending upon the timescale of operation of the generators, distributed energy resources, energy storage systems and DR programs, the obtained schedule is applied and these components are controlled.

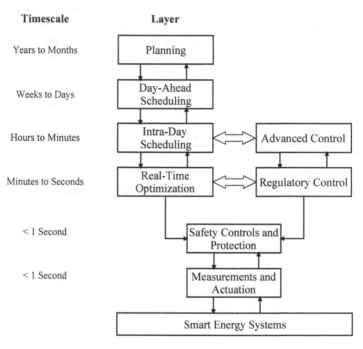

**Figure 1.2**  Layers of energy system operations

Accordingly, the system performance is essentially governed by two subsystems – a computational subsystem which is responsible for the decision-making process, and a control subsystem which is responsible for physical control and actuation based upon the optimal decisions. The computational subsystem may be composed of a computing platform such as a microcontroller, computer or a web application. This subsystem executes a program which is generally based upon optimization of an energy system operation, and computes an optimal decision. The optimal decision is communicated to the control hardware through fast and reliable communication infrastructure. The control hardware, upon receiving the optimal decision, generates control signals for actual control of the physical devices/equipment.

The computational subsystem in the scheduling layer computes optimal decision in the period of few weeks to minutes. The scheduling decision which is obtained until the day of actual operation is referred to as "day-ahead scheduling". Day-ahead scheduling is performed several times and the obtained schedules are updated until the day of actual operation. The scheduling process which is performed during the day of actual operation is referred to as "intra-day scheduling". Intra-day scheduling can further be categorized based upon smaller intervals as block-ahead (several hours), hour-ahead and intra-hour. Intra-hour schedules may be obtained for periods of 15 minutes to 5 minutes, or even lower, up to a minute. However, the time-ahead decisions are not well suited to real-time operating conditions, since they are characterized by uncertainties which are not accurately forecast much ahead of time. Thus, real-time optimization is performed much closer to actual time of operation, in the timescale of minutes to seconds.

Microgrids are composed of multiple system components which have lower response time, such as diesel generators, energy storage systems and loads. The control of these multiple parallel operating devices in energy systems is obtained through the control subsystem. Optimal schedule decisions are implemented through advanced control systems in the timescale of hours to minutes, which ensure joint operation of multiple system components. It is worth noting, that real-time scheduling which is designed to track uncertainties occurring in the system in the timescale of minutes to seconds, may be constrained by computational time in certain scenarios or in particular operations. Hence, for the purpose of second-to-second tracking and control of the system, regulatory control and actuation systems are designed. These are designed to control individual devices within the time frame of a few seconds. Further, the safety, protection and measurement are performed by the control subsystem at intervals of less than a second.

### 1.3.2 ENERGY MANAGEMENT SYSTEMS

Performance of energy systems is dependent upon the efficiency of the computational and control subsystems, and coordination of its various components for the multiple system operations. In the perspective of microgrids, system operations require efficient control of the multiple components such as diesel generators, renewable energy sources, fuel cells, battery energy storage systems and loads. The need to coordinate several resources with the stipulated time frame as required by the respective system operation envisages the need for a system operator. Accordingly, optimized operation of the microgrid is achieved by efficient coordination between well-designed control and computational subsystems, through an energy management system (EMS). The core purpose served with an EMS is the improvement of energy efficiency of a system, through coordination mechanisms to balance the generation and consumption systems.

---

### Energy Management Systems

According to ISO 50001, EMS is defined as "the entity of inter-relative or interactive elements for the implementation of an energy policy and strategic energy goals as well as processes and methods for the attainment of these goals" [16]. However, in perspective of energy systems, the definition of EMS is more closely related to VDI standard VDI-4602 which defines energy management as *"the forward-looking, organised and systematic coordination of the procurement, conversion, distribution and utilization of energy in order to cover requirements and which takes ecological and economic objectives into consideration"* [36].

The computational subsystem essentially performs the task of determining the most optimal point of operation of the system and its various components. This requires well-designed and robust algorithms and solution techniques, which are capable of computing most appropriate dispatch schedules and control set points, in dynamic environments. From the perspective of microgrids, the system may be subjected to a number of phyical constraints pertaining to the technical limits of the generating and storage units, DR constraints which may be relating to the appliance operation and/or users' comfort constraints, as well as network (nodal voltage, thermal limits, etc.) constraints. This implies that the computational subsystem of an EMS is essentially an *optimizer*, wherein the algorithm may be developed as a single-objective or multi-objective optimization problem. The optimizer for energy systems may be designed in view of the following major objectives.

1. *Economical Objectives:* These include minimization of overall operational costs including fuel costs, operation and maintenance costs, start-up costs associated with generating units such as cold start and hot start costs and equipment costs such as capital investment and replacements.
2. *Environmental Objectives:* These include objectives to minimize adverse effects on the environment through minimization of carbon emissions and other greenhouse gases from the generating units and reduction of the total carbon footprint of the system.
3. *Load Profiling Objectives:* These include load-shaping objectives of the utility which aim toward minimizing adverse effects on the distribution system components, and thus, defer investments on utility assets. The various objectives under load profiling are collectively referred to as demand side management techniques, which include load shifting, peak clipping, valley filling, etc.

## 1.3.3   DESIGN AND FORMULATION OF OPTIMIZER MODEL

The foremost requirement for designing any process of an energy system operation is to identify the frequency and timescale of executing the operation, availability of measurement data and forecast of input variables. Accordingly, an operation may be identified as an *online* or *offline* problem. An online operation is one where the information required for computation is revealed incrementally, and decisions are required to be made before complete availability of the input data. In contrast, problems wherein all the data is assumed to be known completely when the computation is performed are called offline problems. From the perspective of microgrids, system designing, long-term and short-term planning, operational planning and time-ahead scheduling are some of the common offline problems. Real-time scheduling, energy management and power sharing are some common online problems in energy systems.

Offline problems, by definition, are implemented in the longer timescale and have low computational speed. Since these problems are generally related to the system designing process, they include complicated technical and economic models which increase the dimensions and complexity of the mathematical models. As a consequence, these problems have higher computational burdens and longer solution time. However, since the optimal values of the offline decision variables are not applied in practice during real-time operation, the higher computational time is generally acceptable. On the contrary, online problems in smart energy systems require computation of decisions during real-time operation for optimal control of the multiple fast-acting devices and uncertain RES.

At the core of an optimizer is an optimization problem which executes in a small timescale and computes optimized decision variables for generation of the control signals of individual system controllers. The general approach for designing an optimizer is composed of three steps. The foremost step is identification and formalization of the objectives of the desired EMS. Further, all essential technical constraints must be identified. Secondly, mathematical expressions of the optimizer objective function and constraints are formulated. This set of mathematical relationships that completely define the problem under consideration and general/technical essential features for the specific system is referred to as a *model*. A general form of a model may be expressed as an optimization problem, as follows.

$$\min_{x} f(x),\ x \in \Re^n \tag{1.1}$$

where $f(x)$ is the minimization objective of the optimizer, $x$ is a set of decision variables and $\Re$ refers to the Euclidean space. The problem is subjected to the following equality and inequality constraints:

$$g(x) = 0 \tag{1.2}$$

$$h(x) \leq 0 \tag{1.3}$$

At this step, problem formalized in (1.1)-(1.3) is evaluated. Suitable solution algorithms are analyzed and tested for their relative performances in solving the EMS model. Since energy management is usually a small-timescale problem, computational speed is a very important parameter desired for the solution algorithm. Accordingly, if the mathematical problem cannot be solved in the time frame of practical implementation of the EMS, the model is suitably transformed using different distributed optimization or decomposition approaches. The modified optimization model is then executed and optimum values of decision variables are obtained. This constitutes the third step. The model is then tested for performance evaluation under practical operating conditions. Once the model has been tested and analyzed for performance under real-time conditions, it is generally useful to perform *post-optimality* analysis through the evaluation of sensitivities. Sensitivity of one variable $\alpha_i$ to another variable $\alpha_j$, by definition, refers to the change in value of $\alpha_i$ due to a disturbance in the value of $\alpha_j$. This is performed to identify the sensitive variables, whose value cannot be changed without changing the optimum value of the optimization model. It must be noted that developing a suitable EMS model is a trade-off between the design accuracy and computational efficiency.

## 1.3.4   REAL-TIME OPTIMIZATION

Real-time optimization refers to a set of optimization methods that include practical real-time inputs and state estimates in the mathematical optimization model to optimally execute real-time processes, while satisfying certain practical constraints. Small-timescale problems in energy systems such as optimal dispatch of generation sources and energy storage systems, demand response programs, load control and peer-to-peer energy sharing are some online operations which necessitate the development of real-time optimization models in smart energy systems. Real-time implementation of a model is faced with the challenge of uncertainties in measured inputs, and requires re-evaluation of optimum values after regular time intervals. It is often the case that input variables of the objective function and constraints of the optimization model, which are typically measured ahead of time, differ measurably from the forecast values at the actual time of occurrence. As a result, the outputs of the optimization model tend to be sub-optimal, rendering the model inaccurate.

In microgrids, uncertainties arise mainly due to increased penetration of RES, dynamic user preferences and real-time tariff, among other factors. It is noteworthy that solar insolation is highly uncertain, which may result in fluctuation of PV output up to 15% of its rated capacity within 1 minute. Furthermore, wind speed suffers from high temporal variations, as a consequence of which they may rise to 2-3 times of their forecast values within a few seconds. This implies that wind power output may become 8-27 times within a period of a few seconds [29]. Uncertainties in input data may lead to potential inaccuracies of the optimization models. Since, the continuous effort of a system operator is motivated at balancing load and generation, under - or overestimates of the forecasts are detrimental to system reliability. System operators are, thus, forced to curtail loads or pay penalties for the unmet demand due to the sub-optimal schedules.

Smaller energy systems are greatly affected by uncertainties due to lower inertia and limited energy buffers. In case the system is grid-connected, the distribution network is increasingly stressed, while the unmet load increases in offgrid systems, leading to user-dissatisfaction. A prolonged mismatch in the generation and load may result in significant deterioration of system components due to sub-optimal operation under unforeseen circumstances. The power distribution network and associated assets such as distribution transformers may be subjected to significant overloads, which may accelerate equipment ageing, and incur higher investments due to earlier replacements. Additionally, from the user perspective, unforeseen generation shortfall results in reduced utility and comfort, and causes dissatisfaction to the consumer. Ensuring efficient system operation in such uncertain conditions requires updating the optimal schedules after regular time intervals to match the requirements of real-time optimization.

Real-time optimization is performed either by using measurements of input variables, or by the use of forecasts of the input data, ahead of time of actual operation. Methods which do not use measurements include nominal optimization and robust optimization. Among the methods that use measurements, three broad types of methods are known in literature, namely, *(i)* adaptation of process model, *(ii)* adaptation of optimization and *(iii)* direct input adaptation [11]. These methods are explained as follows:

1. *Adaptation of process model*: Measurements can be used to refine the model, which implies that the mathematical expressions of the functions are updated in between consecutive optimization iterations. This adaptation is performed by means of the update of the model parameters. Examples include two-stage or multi-stage processes.
2. *Adaptation of optimization*: Measurements can also be used to modify the optimization problem directly. With these methods, corrective terms for modifying the optimization model are determined. These terms are directly added to the mathematical expressions in the formulation, which are kept unchanged. Examples of this method are bias update and constraint adaptation.
3. *Direct input adaptation:* This class of methods proposes direct modification of the inputs, in a control-inspired manner. However, the main difficulty in this method is to construct a control problem, which has the desired self-optimizing properties. Examples include tracking active constraints and extremum-seeking control.

As discussed above, handling uncertainty occurrence in energy systems is imminent for reliable system performance and user-comfort satisfaction. For online energy system operations, this may be achieved by design of real-time system optimizers. In operations research, uncertainty handling has been discussed by a number of methods such as sensitivity analysis, robust optimization methods, chance-constraints and stochastic programming. Sensitivity analysis method is a post-optimality process and is, thus, unsuitable for online optimization. Another main distinction among uncertainty handling methods is based upon the type of information available about the uncertainty. In case the uncertain variables are observed to follow standard probability distribution functions, they are referred to as probabilistic uncertainties. However, it is more common in practical situations to have non-probabilistic uncertainties. In accordance, Bertsimas et al. presented the basic distinguishing feature between robust and stochastic optimization as follows. "*In robust optimization the decision-maker constructs a solution that is feasible for any realization of the uncertainty in a given set, while stochastic programming is seeking to immunize the solution in some probabilistic sense to stochastic uncertainty*" [7]. Each of these methods have been briefly discussed below.

### 1.3.4.1  Robust Optimization

Robust optimization is a class of optimization techniques that aims at incorporating variable uncertainty in optimization models by assuming they belong within deterministic uncertainty sets. These algorithms do not require explicit probabilistic distributions of the uncertain variables, and the assumption of well-defined uncertainty sets ensures the solution of the model in polynomial time. As mentioned in the preceding subsections, robust optimization does not require data measurements, and works well with time-ahead forecasts. However, it must be noted that robust optimization is essentially a worst-case optimization problem, which implies that it determines the optimal value of the objective function for the worst realization of uncertainty. This results in an overly pessimistic or conservative solution.

Accordingly, as per the first set of solution techniques for real-time algorithms, a two-stage robust optimization model is well suited to online optimization of energy system operations. While the first stage maximizes/minimizes the objective function for worst-case realization of uncertain variables, the second stage is a recourse decision that performs real-time updates in the optimal solution once the precise information about uncertain variables is available. The two-stage robust optimization model may be mathematically represented as a *min-max-min* problem as follows [29].

$$\min_{x} f(x) + \max_{u} \min_{y} d(y) + e(u) \tag{1.4}$$

$$s.t. \ A(x) \geq b \tag{1.5}$$

$$F(x) + G(y) \leq v \tag{1.6}$$

$$H(u,x) + I(y) + J(u) = w; \ u \in U \tag{1.7}$$

Here, the first *min* is the first stage problem that minimizes the objective $f(x)$ by determining the optimal value of the "certain" decision variable $x$. This implies that the optimal decision of this stage is the *here-and-now* decision, as the variable $x$ is not subjected to uncertainties. Accordingly, the optimal value of these variables are determined in the first stage, and remain fixed for the remaining optimization process. The *max* refers to maximization of uncertainty function $e(u)$, indicating a worst-case realization of uncertain variables $u$. Once the worst-case uncertainties are realized, the second *min* function minimizes the second stage objective function $d(y)$ through computation of optimal decision variables $y$. Since the variable $y$ is optimized after occurrence of worst-case uncertainty, these are known as *wait-and-see* decisions.

In robust optimization problems, the uncertain variables are characterized by uncertainty sets or budgets, which are clearly defined by upper and lower limits. The uncertainty budget $U$ of an uncertain variable $u$ is a measure of uncertainty in the forecast data. This implies that having a wider uncertainty set means that the model is robust against a greater level of uncertainties. However, this also leads to higher costs and more conservative solutions. Thus, choice of budget has to be made keeping in view a careful choice between the conservativeness and robustness of the model. A more commonly used polyhedral uncertainty set is defined by the following expressions [42].

$$U = \{u^t \in \mathbb{R}^n : u^{t,min} \leq u^t \leq u^{t,max}, \forall\, t$$

$$\mu^l \leq \frac{\sum\limits_{t \in T} u^t}{\sum\limits_{t \in T} u^{f,t}} \leq \mu^u \} \tag{1.8}$$

where $u^{t,min}$ and $u^{t,max}$ are the minimum and maximum values of the uncertain variable $u$ at time $t$, respectively. Further, $u^{f,t}$ refers to the forecast value of $u$ at time $t$ and $\mu^l$ and $\mu^u$ are the lower and upper limits of the corresponding uncertainty sets.

Solution of a two-stage robust optimization problem requires decomposition techniques which have been widely used in literature. These decomposition approaches such as Bender's decomposition and Column and Constraint Generation have the advantages of higher computation efficiency and lower solution time. It is also noteworthy that the computational time does not increase significantly with an increase in the levels of uncertainty. These characteristics are well suited for online optimization required in energy systems. The decomposition algorithms divide the robust optimization model into a master and a slave problem, which are solved iteratively. Thus, the first stage solution is optimized for the worst-case realization of uncertainty in the second stage. However, in practice, the worst-case uncertainties are rarely encountered. Thus, for online algorithms, *real-time adjustments* of the second stage variables are made after actual occurrence of uncertain variables.

### 1.3.4.2  Stochastic Programming with Recourse

Stochastic programming is the field of operations research that assumes probability distributions can be estimated for the uncertain variables in the optimization model. These probability distribution functions are then employed to generate random scenario occurrences, and an expected value of the objective is optimized. While robust optimization deals with hard constraint satisfaction, stochastic programming is generally used for problems with soft constraints, which are expected to yield good results on average. The stochastic programming with recourse is essentially a two-stage problem which performs the first stage optimization in the absence of uncertainties, and delays the second stage decision to a later time when uncertainties are actually present or a precise information is available for the uncertain variables.

Let us consider an energy system operation where certain information is available time-ahead, and the uncertainties are revealed closer to real-time. A stochastic optimization model can be formulated for this operation, where the first stage problem is $min\, f(x) : x \in F$ and the second stage problem is $min\, g(y) : y \in G(x,u)$. Here, $x$ is the "certain" decision variable of the first stage. It is known with certainty, hence, the first stage decision can be computed before actual occurrence of uncertainty. Accordingly, these variables are referred to as "here and now" decision variables. Upon occurrence of uncertainty $u$ during real-time operation, the second-stage optimization is performed where the uncertainty scenarios are determined, each associated with an occurrence probability $\lambda$. The decision variables of this stage $y$ are accordingly referred to as recourse or "wait and see" decision, and update the first stage decision. Accordingly, the two-stage stochastic programming model can be mathematically represented as follows [30].

$$min \ \{f(x) + \mathbb{E}[H(x,u)]\} \qquad (1.9)$$

Here, the second stage optimization decision, also referred to as the recourse decision, is represented by $H(x,u)$. The expected value of the second stage optimization over a period of time is given by $\mathbb{E}[H(x,u)]$. For the ease of computation, the stochastic model is reformulated as a deterministic equivalent model. However, this reformulation is associated with the assumption of a finite number of occurrences of uncertainty scenarios. Each of these uncertainty scenarios is associated with a probability of occurrence. Accordingly, expected value of the second stage can be expressed as follows [39].

$$\mathbb{E}[H(x,u)] = \sum_{n=1}^{N} \lambda_n H(x,u_n) \qquad (1.10)$$

The general solution process of a stochastic programming model with recourse includes the determination of uncertainty scenarios based upon probability distribution functions of the uncertain variables. From the perspective of microgrids, the commonly present uncertain variables are PV output, wind power and electrical load. These three uncertain variables have been observed to follow the *Beta, Weibull* and *Normal* distributions, respectively. The scenarios are generated using sampling techniques such as Monte Carlo Simulation. However, the high computational burden of executing the operation for each sample renders the model unsuitable for online optimization. Thus, scenario reduction techniques such as Forward Reduction and Backward Reduction have been found effective in reducing the large number of uncertainty scenarios to a small, yet representative set. Accordingly, an equivalent deterministic model is obtained which is then solved using classical solution techniques. The deterministic equivalent model of the two-stage stochastic programming framework can be expressed as follows.

$$\min_{x,y_n} \ \{f(x) + \sum_{n=1}^{N} \lambda_n g(y_n)\}; \ x \in F; \ y_n \in G(x,u_n) \qquad (1.11)$$

Here, $N$ refers to the number of uncertain scenarios, each associated with an occurrence probability $\lambda_n$.

### 1.3.4.3 Chance-Constrained Optimization

While robust optimization deals with problems with deterministic uncertainty sets, there exist several such variables whose values cannot be defined by strict boundaries. Examples of such uncertain variables are those whose probability distributions have long tails. A robust optimization approach to solve such models would be to consider very large values for the bounds of such a variable. This, however, would lead to overly conservative solutions. An alternative and more effective solution to optimize models with uncertain parameters having long tails is chance-constrained optimization. Chance-constraints ensure the probability of satisfaction of particular constraints, in contrast to robust optimization which handles all constraints as hard constraints.

Consider a general constraint, where $b_i$ is an uncertain variable which follows normal distribution function. The optimization model with the original constraint may be mathematically represented as follows [13].

$$\min c^T x \tag{1.12}$$

$$\sum_{j=1}^{n} a_{ij}x_j \leq b_i \tag{1.13}$$

The corresponding chance constraint satisfies a very high probability $\gamma$ with which the original constraint must be satisfied. Thus, the chance constraint may be mathematically be represented as follows.

$$P\left\{ \sum_{j=1}^{n} a_{ij}x_j \leq b_i \right\} \geq \gamma \tag{1.14}$$

The deterministic equivalent of this chance constraint may be expressed as follows.

$$\sum_{j=1}^{n} a_{ij}x_i \leq \mu_i + K_\gamma \sigma_i \tag{1.15}$$

where $K_\gamma$ is a statistical constant and $\mu_i$ and $\sigma_i$ are the mean and standard distribution of the normal distribution function, respectively. Thus, in optimization models that deal with soft constraints, chance constraints offer higher guarantee of constraint satisfaction. Accordingly, deterministic equivalents of chance constraints may be included in the stochastic optimization model to offer a higher degree of reliability.

### 1.3.5 BIG DATA ANALYTICS

The transition from traditional power systems to smart energy systems is associated with the emergence of a communication layer which facilitates large exchange of data among several sensors, smart meters, monitoring equipment and management systems. The characteristic functionalities of smart energy systems such as self-healing and restoration, reconfiguration, autonomous control, demand response and transactive energy require bidirectional flow of power and information, over various timescales. The bidirectional flow of power may occur between different consumers, wherein certain consumers are capable of power generation due to the integration of distributed energy resources such as PV systems, wind turbine generators, fuel cells, energy storage systems, and are more specifically referred to as *prosumers*. Power may also flow between the prosumers and the utility grid, prosumers to energy storage systems, microgrid to microgrid, apart from the conventional flow of power. Similarly, bidirectional flow of information occurs between the system operators and aggregators, aggregators and local energy management systems, which further communicate with controllers of the distributed generation sources, battery charge controllers and load aggregators and individual smart plugs. These operations require huge data collection, storage, processing and analysis which may result in the generation of significant amounts of data every year.

Traditional data handling approaches are inadequate and inefficient to cope with the unprecedented high frequency and volume of data being generated continuously in the modern energy systems [20]. The smart grid structure is very complex and has close interdependence and interactions among power system components, information and communication technology, internet of things, and human interaction. Recent advances in the smart grid aim at integration of multiple energy sources for interlinking electrical, thermal and gas networks to achieve higher system efficiency, thus, necessitating generation of an enormous amount of raw data. Thus, big data in smart energy systems are a heterogeneous set of information which are available in different formats, at different frequencies of time, generated at different locations, are mostly asynchronous and are frequently subjected to noise and uncertainties [14]. Owing to the huge quantities of data, computational time and space is significantly increased, rendering them unsuitable for real-time optimization processes.

Big data analytics, which refers to the collective process of handling large volumes of data, discovering useful information from the data and processing it to obtain meaningful control inputs and decisions, appears as a timely and suitable solution to this problem faced by modern energy systems. This data consists of significant amount of useful information which can assist efficient operation of real-time microgrid operations and ensure secure and stable power networks. Further, there have recently been increasing concerns over system security and privacy with the widespread integration of communication infrastructure. Reliable storage and processing of the overwhelming volume of heterogeneous data will further complicate the critical system operations, as dimensions of the energy systems increase in size and functionality over time. Thus, big data, in coordination with machine learning and artificial intelligence (AI), is expected to open up new avenues which promise revolutionary transformations to the conventional energy system operation.

Big data in smart energy systems holds its basic five characteristics, known as the universal 5V big data model, which have been explained as follows [44].

1. *Volume:* refers to the significant amount of data that is generated in the distribution systems or smart microgrids by the numerous smart meters and sensors. This envisages the need for large numbers of safe and reliable data sets. A suitable solution for handling huge data volumes is to promote distributed architectures, which prevent long transmissions of data.

2. *Velocity:* refers to the speed at which the smart meters, sensors and other devices generate the huge volumes of data in energy systems. The increasing need for online processes requires utilization of high volumes of data at high speed. Considering that a smart meter generates data every 15 minutes,

1 million devices would generate 35.04 billion data entries, accounting to 2920 Terabytes each year [14].

3. *Variety:* refers to the heterogeneous nature of data flowing in smart energy systems. The large variety in data occurs due to multiple distributed sources of data, generating multi-dimensional data in different formats and structures. Smart energy systems, as discussed above, witness different types of structured, semi-structured and unstructured data.

4. *Veracity:* refers to the quality and accuracy of data. Data is always subjected to noise during transmission in the network. This can lead to erroneous outputs if the data is not processed suitably. Thus, data pre-processing and state estimation is always the prerequisite for data utilization. In smart energy systems, stability of system operation depends on reliability of the data.

5. *Value:* refers to the process of extracting useful and valuable information stored in the data. It must be noted that with increase in the volume of data, the density of valuable information is reduced. In the smart energy system paradigm, intelligent devices and applications like forecasting models utilize data to forecast future load and generation.

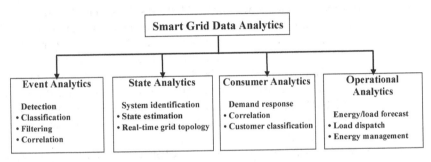

**Figure 1.3**  Application of big data analytics in smart energy systems [8]

Big data analytics finds several applications in smart energy systems and market operations. These include fault detection, transient stability analysis, state estimation, power quality monitoring, renewable energy generation and load forecasting, load profiling and load disaggregation, among other applications. Figure 1.3 lists down the various applications of data analytics to smart energy systems. However, deployment of big data analytics in smart energy systems poses certain challenges. A brief review of some important challenges faced by energy management systems is discussed below [8].

1. *Data Volume*: High volumes of data flow require increased number of data storage resources and computational requirements. Potential solution to this challenge is the use of distributed computing, parallel processing and cloud computing for reduction of dimensionality of energy system data.

2. *Data Security*: Increased functionality imparted to prosumers comes with the need to secure their privacy. Smart meter data, if not handled properly,

can compromise users' privacy and integrity. This necessitates the need for data protection through encryption, anonymization and aggregation, in various applications.

3. *Data Uncertainty*: From the perspective of microgrids, RES and load forecast information are subjected to significant amount of uncertainty which adversely affects decision-making processes. Probabilistic data analytics, stochastic processes and data mining techniques can be incorporated for removing noise, inconsistencies and redundancies in data.

4. *Time Synchronization*: Response time of different components and devices vary. The asynchronous nature of data poses significant challenges for real-time operations in smart energy systems. Unsynchronized data may lead to sub-optimal decisions, which may risk system stability and security. Power system operations utilize synchrophasors which provide time synchronized data based on radio clocks or satellite receivers.

## 1.3.6   ENERGY BLOCKCHAIN

Big data analytics require suitable web-based platforms for secure data storage and processing. Increasing need for real-time operations in modern energy systems is bound to incur significant costs for data handling. Blockchain is a shared ledger that has emerged as a solution for efficient online data storage. It serves the two main requirements of big data analytics viz. data security and preserving its *value*. Blockchain is essentially a sequence of blocks that store information about successive operations which take place. It facilitates transparent operations without the involvement of intermediaries. The security and autonomy of a blockchain are ensured by smart contracts and consensus mechanism [1].

In the context of smart energy systems, blockchain is referred to as *energy blockchain*, and is mainly used in the following applications, namely peer-to-peer energy trading, microgrid operations and cyber physical security [18, 19]. Considering the example of energy trading, two individual participants can carry out a transaction through smart contracts, without the involvement of a third party. A smart contract is integrated in the blockchain and executed as a software agent upon satisfaction of certain preconditions. Various software platforms are available for formulating smart contracts, such as Ethereum Blockchain and Hyperledger Project.

The three main characteristics of a blockchain are decentralization, transparency and immutability. The important features of a blockchain which make it suitable for real-time transative operations are as follows [22].

1. Consensus mechanism which allows necessary agreement of all the participant nodes on each operation ensures the integrity of data stored in the blockchain.

2. Communication requirements are minimized since each node directly communicates only with the blockchain.

3. It is a completely distributed framework which does not require the involvement of a central controlling entity, which might lead to unnecessary communication costs as well as time-delays .

4. The risk of single point of failure or single points of compromise is eliminated due to the distributed architecture, wherein nodes can directly communicate to the network through individual interfaces.

5. Blockchain provides privacy and immutability to the data, which refers to inability to alter transactions that have already been completed. Each block is hashed using cryptographic hashing functions, based upon its data contents and the hash of its previous block, which ensures blockchain security.

## 1.4  INTELLIGENT CLOUD-BASED GRID APPLICATIONS

Energy systems are composed of multiple controllable resources including distributed generation sources such as RES and fuel cells, energy storage systems such as batteries and supercapacitors, along with responsive loads. Energy system operations require several coordinated control actions which are governed by physical limitations on the system such as thermal limits of the lines and bus voltage limits, while ensuring real and reactive power balance at all buses. These control actions such as load scheduling, economic dispatch, fault clearing, re-synchronization and islanding are coordinated over different timescales, from a few seconds to hours. Further, energy and power trading in electrical networks spans over a long duration, starting with long-term contracts signed over a period of months to years, day-ahead markets which compute schedules up to the actual day of operation, intra-day trading, and finally real-time power exchanges.

It is also noteworthy, that the agents engaged in these operations are spatially distributed and located at different locations in the energy systems. Further, the environment around each of these agents may be unique and dynamic, which might deviate from the design conditions of the controllers. It is, thus, evident that the planning, scheduling, operation and control processes in an energy system are spanned over the timescale of years to less than a second, over wide areas, and under varying conditions, thus, highlighting the need for temporal as well as spatial coordination framework. Accordingly, three major control strategies have been widely employed in the control of energy systems, namely, *(a)* centralized, *(b)* decentralized and *(c)* distributed control [31].

### 1.4.1  CENTRALIZED CONTROL

Centralized control refers to the control architecture where there is a single decision making entity. The different system components are controlled by a central controlling agent. This agent is responsible for monitoring, collecting and analyzing the data during real-time system operations. The centralized approach has been widely used in practice and has gradually evolved over the years. Common applications of centralized control in power systems are microgrid control, economic dispatch, unit commitment, wholesale electricity markets, energy management systems, etc. However, over the years, it has been realized that with gradual development of active distribution systems which are composed of distributed energy resources, controllable energy storage systems, flexible loads, the dimensions of control variables are increasing. This significantly increases the complexity of an optimization or control model. This leads to very high computational time which is unsuited for real-time operations. Thus, for smart energy systems, centralized control doesn't appear as a suitable technique due to the following major reasons.

1. The computational burden increases with greater number of control variables. Additionally, the problem may become infeasible as the dimension increases.
2. Transmission of information may become increasingly challenging as and when the area under control increases.
3. Centralized control systems suffer from the problem of single point of failure, rendering this type of control unsuitable for larger control system where the number of entities affected from failure is high.
4. The topology of such systems cannot be easily modified or expanded without affecting other system components, which is a basic requirement for modular microgrids.
5. Centralized control of a system requires an extensive communication network to control system components.

## 1.4.2  DECENTRALIZED CONTROL

Decentralized control refers to the type of control systems where instead of a single central controller, multiple localized controllers perform decision making in an autonomous and independent manner. The local agents use localized information for decision making, and do not depend upon the central control entity. The decision of local agents depends upon local measurement of input variables, which minimizes the requirement for an extensive communication network, as required by centralized control. This also minimizes the risk of single point of failure, and interrupted system operation due to communication delays. However, this form of control does not guarantee global optima due to absence of communication and information exchange among agents. Thus, decentralized control is prone to risks of being trapped in local optima and giving sub-optimal solutions.

## 1.4.3  DISTRIBUTED CONTROL

Distributed control systems comprises of multiple localized agents which use local measurements and information shared by other agents for performing the decision making process. The ability to communicate and share information among agents guarantees a global optimum for system operation. However, this creates concerns for data privacy. A number of communication protocols with data privacy standards and encryption algorithms are being developed. Example of this type of control is the energy blockchain, discussed in the previous sections. Distributed systems may be categorized into three major types: *(i)* decomposition, *(ii)* networked optimization and *(iii)* non-cooperative games. A broad distinction between these types is

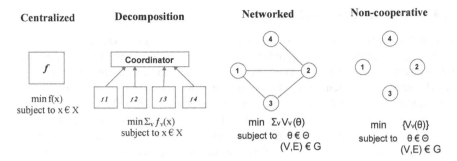

**Figure 1.4** Illustration of centralized optimization, decomposition, networked optimization and non-cooperative optimization [41]

illustrated in Figure 1.4. Decomposition refers to breakdown of a complex objective into multiple smaller problems that are individually solved. Networked optimization includes systems where each agent can communicate only with the immediate neighbors. This requires optimization algorithms wherein the communication structure is included in the model. Non-cooperative games include problems where the agents do not coordinate with each other, thus eliminating the requirement for complex coordination protocols and communication infrastructure.

Owing to the specific characteristics of distributed control approaches, the requirements of optimization algorithms for these models can be categorized into the following two types.

1. Lagrangian decomposition based methods, which include dual decomposition and alternating direction method of multipliers.
2. Methods based on Karush–Kuhn–Tucker necessary conditions.

The various applications of smart power systems which require decentralized control are distribution system operations including optimal power flow, demand side management, peer-to-peer energy trading, demand response and stability studies. The various features of distributed control which favors their application in smart energy systems are listed as follows.

1. Computational burden is distributed over a number of agents. This results in multiple parallel processes, thereby, reducing the computation time.
2. The networks are immune to single point of failure, and hence have higher reliability.
3. An important characteristic of smart energy systems is its modularity, which is well supported by distributed control.
4. Distributed control systems support dynamic network topology which is required by several transactive energy frameworks.

## 1.4.4  MULTI-AGENT SYSTEMS

Multi-timescale variation of power flows, RES intermittency and variability, network reconfiguration ability and plug-and-play characteristics highlight the dynamic nature of smart energy systems. Secure and reliable operation of these systems envisages the need for efficient computation and real-time control. Accordingly, AI has been identified as a suitable solution for online operations. Russell and Norving have introduced the concept of *agents*, and defined AI as "the study of agents that exist in an environment and perceive and act" [28]. Appropriately, agents may be thought of as intelligent entities, capable of performing autonomous actions under varying environmental conditions. Although agent-based technology has been in existence for many years, there lacks a uniformly accepted common definition. A common notion that has been identified as the main characteristic of agents is their autonomy and pro-activity. A widely used definition was proposed by Wooldridge and defines an agent as "a software (or hardware) entity that is situated in some environment and is able to autonomously react to changes in that environment" [38].

The main force behind the development of agent-based technology is the fact that it supports distributed control architecture, necessary for most modern day problems. By definition, multi-agent systems consist of groups of agents that co-exist, communicate, and coordinate with each-other. Multi-agent systems enable multiple parallel processes by decomposing a complex objective into several simpler objectives, each assigned to an independent intelligent agent. This significantly reduces computational time and eliminates single point of failure, which are prerequisites for online control. Multi-agent systems differ from traditional AI techniques due to the fact that most conventional AI techniques are designed for static environments. This is due to the ease and simplicity of design that static environments offer, which enables easier mathematical formulation. However, smart energy systems are decentralized, market-oriented, multi-variable and complex. Further, the increasing number of online problems in modern distribution networks envisages the need to develop suitable control techniques such as multi-agent systems, which offer autonomous and proactive control in dynamic environments. This enables highly networked problems in smart energy systems, which are executed in real-time.

## 1.5  CONCLUSION

Owing to ever-increasing energy demands and rising concerns over global warming and climate change, urban settlements are undergoing rapid structural and technological changes, and popularly being referred to as smart cities. This has been accompanied by widespread integration of RES in the energy systems which are highly intermittent and necessitate the development of advanced controls. In order to impart flexibility to RES-penetrated energy systems, energy storage and flexible demand have been identified as potential solutions. Due to the diverse spatial distribution of generation, storage and load resources in the electrical distribution grid, distributed control architecture is required to ensure energy demand-supply balance at all times at an acceptable reliability level. Consequently, smaller and smarter distributed energy systems such as microgrids and smart buildings have been recognized as potentially active resources with demand response capabilities to assist grid operations. In view of this, this chapter discusses the dynamics of energy system operations, DR and DSM strategies, and associated challenges. Further, since these programs requires high computational efficiency to match real-time grid requirements, this chapter discusses design and development of suitable control algorithms and uncertainty mitigation techniques for real-time operation. Moreover, online system operation requires high-speed data transfer among system components, which raises challenges of data processing and handling, with small-lead time operations. Accordingly, this chapter further discusses big-data analytics and blockchain technology for real-time energy system operations. The chapter concludes by discussing the cloud-based implementation architectures of smart energy systems for autonomous and intelligent operations in real-time.

## REFERENCES

1. Shubhani Aggarwal, Rajat Chaudhary, Gagangeet Singh Aujla, Anish Jindal, Amit Dua, and Neeraj Kumar. Energychain: Enabling energy trading for smart homes using blockchains in smart grid ecosystem. In *Proceedings of the 1st ACM MobiHoc Workshop on Networking and Cybersecurity for Smart Cities*, pages 1–6, 2018.

2. Maha Al-Zu'bi and Vesela Radovic. *References', SDG11 sustainable cities and communities: Towards inclusive, safe, and resilient settlements (Concise guides to the United Nations sustainable development goals).* Emerald Publishing Limited, 2018.

3. Gagangeet Singh Aujla, Sahil Garg, Shalini Batra, Neeraj Kumar, Ilsun You, and Vishal Sharma. Drops: A demand response optimization scheme in sdn-enabled smart energy ecosystem. *Information Sciences*, 476:453–473, 2019.

4. Gagangeet Singh Aujla and Neeraj Kumar. Mensus: An efficient scheme for energy management with sustainability of cloud data centers in edge–cloud environment. *Future Generation Computer Systems*, 86:1279–1300, 2018.

5. Gagangeet Singh Aujla and Neeraj Kumar. Sdn-based energy management scheme for sustainability of data centers: An analysis on renewable energy sources and electric vehicles participation. *Journal of Parallel and Distributed Computing*, 117:228–245, 2018.

6. Gagangeet Singh Aujla, Mukesh Singh, Neeraj Kumar, and Albert Zomaya. Stackelberg game for energy-aware resource allocation to sustain data centers using res. *IEEE Transactions on Cloud Computing*, vol. 7, no. 4, pp. 1109–1123, 2017.

7. Dimitris Bertsimas, David B Brown, and Constantine Caramanis. Theory and applications of robust optimization. *SIAM Review*, 53(3):464–501, 2011.

8. Bishnu P Bhattarai, Sumit Paudyal, Yusheng Luo, Manish Mohanpurkar, Kwok Cheung, Reinaldo Tonkoski, Rob Hovsapian, Kurt S Myers, Rui Zhang, Power Zhao, et al. Big data analytics in smart grids: State-of-the-art, challenges, opportunities, and future directions. *IET Smart Grid*, 2(2):141–154, 2019.

9. John Conti, Paul Holtberg, Jim Diefenderfer, Angelina LaRose, James T Turnure, and Lynn Westfall. International energy outlook 2016, with projections to 2040. May 2016. Washington, DC, USA: US Energy Information Administration (EIA). doi: DOE/EIA-0484, 2014.

10. US DOE. Department of energy, office of electricity delivery and energy reliability. In *Office of Electricity Delivery and Energy Reliability Smart Grid R&D Program: Summary: Report: 2012 DOE Microgrid Workshop–July*, pages 30–31, 2012.
11. Grégory François and Dominique Bonvin. Real-time optimization: Optimizing the operation of energy systems in the presence of uncertainty and disturbances. *Proceedings of the 13th SET Conference*, page E40137, 2014.
12. CIGRÈ C6.22 Working Group. Department of energy, office of electricity delivery and energy reliability. In *Microgrids 1: Engineering, Economics, & Experience – Capacibilities, Benefits, Business Opportunities, and Examples – Mircogrids Evolution Roadmap*, 2015.
13. Frederick S Hillier. *Introduction to operations research*. Tata McGraw-Hill Education, 2012.
14. Machine Learning IEEE Smart Grid Big Data Analytics and Artificial Intelligence in the Smart Grid Working Group. Big data analytics in the smart grid. 2017.
15. International Renewable Energy Agency (IRENA). Off-grid renewable energy systems:status and methodological issues. 2015.
16. ISO. 50001: 2018. *Energy Management Systems – Requirements with Guidance for Use*.
17. Anish Jindal. *Data Analytics of Smart Grid Environment for Efficient Management of Demand Response*. PhD thesis, Thapar Institute of Engineering and Technology Patiala, 2018.
18. Anish Jindal, Gagangeet Singh Aujla, and Neeraj Kumar. Survivor: A blockchain based edge-as-a-service framework for secure energy trading in sdn-enabled vehicle-to-grid environment. *Computer Networks*, 153:36–48, 2019.
19. Anish Jindal, Gagangeet Singh Singh Aujla, Neeraj Kumar, and Massimo Villari. Guardian: Blockchain-based secure demand response management in smart grid system. *IEEE Transactions on Services Computing*, 13(4):613–624, 2020.
20. Anish Jindal, Neeraj Kumar, and Mukesh Singh. A unified framework for big data acquisition, storage, and analytics for demand response management in smart cities. *Future Generation Computer Systems*, 108:921–934, 2020.
21. Anish Jindal, Mukesh Singh, and Neeraj Kumar. Consumption-aware data analytical demand response scheme for peak load reduction in smart grid. *IEEE Transactions on Industrial Electronics*, 65(11):8993–9004, 2018.
22. Olamide Jogunola, Mohammad Hammoudeh, Bamidele Adebisi, and Kelvin Anoh. Demonstrating blockchain-enabled peer-to-peer energy trading and sharing. In *2019 IEEE Canadian Conference of Electrical and Computer Engineering (CCECE)*, pages 1–4. IEEE, 2019.
23. Devinder Kaur, Gagangeet Singh Aujla, Neeraj Kumar, Albert Y Zomaya, Charith Perera, and Rajiv Ranjan. Tensor-based big data management scheme for dimensionality reduction problem in smart grid systems: Sdn perspective. *IEEE Transactions on Knowledge and Data Engineering*, 30(10):1985–1998, 2018.
24. Nandkishor Kinhekar, Narayana Prasad Padhy, Furong Li, and Hari Om Gupta. Utility oriented demand side management using smart ac and micro dc grid cooperative. *IEEE Transactions on Power Systems*, 31(2):1151–1160, 2015.
25. Chaojie Li, Xinghuo Yu, Wenwu Yu, Guo Chen, and Jianhui Wang. Efficient computation for sparse load shifting in demand side management. *IEEE Transactions on Smart Grid*, 8(1):250–261, 2016.

26. Thillainathan Logenthiran, Dipti Srinivasan, and Tan Zong Shun. Demand side management in smart grid using heuristic optimization. *IEEE transactions on smart grid*, 3(3):1244–1252, 2012.
27. Henrik Lund, Poul Alberg Østergaard, David Connolly, and Brian Vad Mathiesen. Smart energy and smart energy systems. *Energy*, 137:556–565, 2017.
28. P Russel Norvig and S Artificial Intelligence. *A Modern Approach*. Prentice Hall, 2002.
29. S Sharma, A Verma, Y Xu, and BK Panigrahi. Robustly coordinated bi-level energy management of a multi-energy building under multiple uncertainties. *IEEE Transactions on Sustainable Energy*, pages 1–1, 2019.
30. S Sharma, Y Xu, A Verma, and BK Panigrahi. Time-coordinated multienergy management of smart buildings under uncertainties. *IEEE Transactions on Industrial Informatics*, 15(8):4788–4798, 2019.
31. Wencong Su and Alex Huang. *The Energy Internet: An Open Energy Platform to Transform Legacy Power Systems Into Open Innovation and Global Economic Engines.* Woodhead Publishing, 2018.
32. Bernard Tenenbaum, Chris Greacen, Tilak Siyambalapitiya, and James Knuckles. *From the bottom up: How small power producers and mini-grids can deliver electrification and renewable energy in Africa.* The World Bank, 2014.
33. United Nations. About the Sustainable Development Goals: https://www.un.org/sustainabledevelopment/sustainable-development-goals/.
34. United Nations. Sustainable Development Goal 11: https://sustainabledevelopment.un.org/sdg11.
35. United Nations Development Programme. Sustainable Development Goals: https://www.undp.org/content/undp/en/home/sustainable-development-goals/background/.
36. VDI VDI-Guideline. 4602, 2007.
37. Catharine Way. *The millennium development goals report 2015*. UN, 2015.
38. Michael Wooldridge. *An introduction to multiagent systems*. John Wiley & Sons, 2009.
39. Y Xu, ZY Dong, R Zhang, and DJ Hill. Multi-timescale coordinated voltage/var control of high renewable-penetrated distribution systems. *IEEE Transactions on Power Systems*, 32(6):4398–4408, 2017.
40. Yan Xu, Zhao Yang Dong, Kit Po Wong, Evan Liu, and Benjamin Yue. Optimal capacitor placement to distribution transformers for power loss reduction in radial distribution systems. *IEEE Transactions on Power Systems*, 28(4):4072–4079, 2013.
41. Bo Yang and Mikael Johansson. Distributed optimization and games: A tutorial overview. In *Networked Control Systems*, pages 109–148. Springer, 2010.
42. C Zhang, Y Xu, and ZY Dong. Robustly coordinated operation of a multi-energy microgrid in grid-connected and islanded modes under uncertainties. *IEEE Transactions on Sustainable Energy*, 11(2):640–651, 2020.
43. Qin Zhang and Juan Li. Demand response in electricity markets: A review. In *2012 9th International Conference on the European Energy Market*, pages 1–8. IEEE, 2012.
44. Yang Zhang, Tao Huang, and Ettore Francesco Bompard. Big data analytics in smart grids: A review. *Energy Informatics*, 1(1):8, 2018.

# Section II

---

## Smart Grids

# 2 Conventional Power Grid to Smart Grid

*Dristi Datta*
Varendra University, Rajshahi, Bangladesh

*Nurul I. Sarkar*
Auckland University of Technology, Auckland, New Zealand

## CONTENTS

## 2.1   INTRODUCTION

The journey of electricity begins with human civilization and it is considered one of the most important innovations in the civilized world as the fire was that of ancient times. Electricity is one of our lifeline items and it can be very hard for us to live in a modern society with electricity. Therefore, there is no other option that we cannot ignore the need for electricity and its importance in our daily life. We are used to electricity in such a way that without it, even a day, it becomes quite impossible for us to stay and lead our life. In the 19th century, the energy was generally cultivated from coal, furnish oil, gas, and various types of fossil fuel which was sufficient at that time. However, with the increasing world population, it will significantly increase the demand for electricity and fossil fuel. If this trend of high consumption of fuel continues in the next decade the world will face a huge shortage of fossil fuel. Therefore, with the increasing demand for electricity in recent years, it is quite impossible to meet up the demand with the traditional grid system [13].

In this context, the fast exhaustion of petroleum product assets and ecological concern has given a consciousness of the generation of renewable energy resources. Therefore, many researchers suggested cultivating more and more energy from green sources for example solar, wind, and biogas.

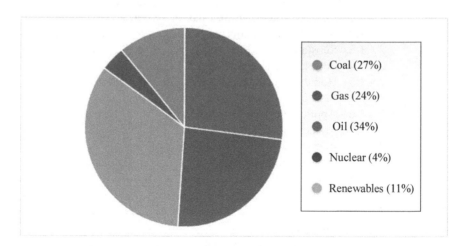

**Figure 2.1**   World total primary energy consumption by fuel in 2018 [2]

According to the current energy consumption as shown in Figure 2.1, about 85% of the all-out vitality is delivered from petroleum products, 4% is created in nuclear plants, and the remaining 11% originates from sustainable sources (essentially hydro, sunlight based, biomass, and wind power) [2].

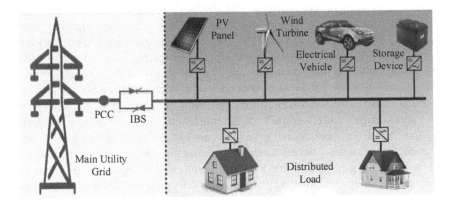

**Figure 2.2**   An illustration of islanded microgrid scheme

A few decades back, the integration with renewable energy with the existing traditional grid was not an easy task as the power that is generated from renewable suffer
from stability problems [10, 7]. For instance, the intensity of the sun varies within a
day that can produce variable DC power only; whereas, the flow of wind also fluctu-
ates in nature and is unpredictable. Therefore, the generated power is variable, and it
needs to be stabilized before connecting the load. Many researchers suggested that to
use storage devices (battery) to store the energy. After that energy can be converted
from DC to AC using inverter to supply to the grid. This method of power generation
is expensive due to the costs of battery and inverter circuits. In some remote areas,
where the grid connection is difficult to reach, a direct DC power system is used in
residential houses for operating light loads such as fans and lights. From this, the
concept of microgrid is induced.

The microgrid is a localized and isolated grid that can operate in a small zone
and used for the electricity supply which is generated by the near-source. However,
the idea and definition of microgrid are expanded and updated with the need for the
power grid. Microgrid can be connected with the main grid via point of common
coupling (PPC) and can receive or deliver power when needed. Therefore, microgrid with distributed generation (DG) is considered as an alternative for supplying
uninterrupted power to the load. The microgrid provides higher efficiency, energy
security, reliability, economics saving, and sustainability [14]. The microgrid system
is of interest to researchers because of its ability to run independently [14]. The block
diagram of a microgrid system is shown in Figure 2.2.

The microgrid consists of generators such as wind, photovoltaic panel, PCC, energy storage element, and controllable loads. The contemplated process of microgrid
follows two operation conditions. First one is a grid-connected technique and the
second one is a standalone technique [38].

During grid connected state the distributed energy resources (DER) backs the main grid whereas rest is connected to it at the PCC. In this contour, the frequency and voltage of the system are being determined by the transmission grid and microgrid can deliberate taking power from the main grid. The microgrid can be disjoined from the main grid when a disturbance (fault, power quality, and voltage collapse) occurs. Meanwhile, the stand-alone microgrid mode is activated [34].

The journey of power sectors is not bounded with microgrid, however, researchers are carried their research to convert microgrid into the smart grid. Because power control and management are considered the most challenging issue from the period of the journey of the conventional grid. The conventional grid suffers lots of problems in dealing with the different kinds of load and there is only the basic control is available from the grid side to the distribution end. Power engineers are interested to convert the overall system into an automated system where numbers of the sensor are being used.

For effective power control and management, several pieces of research are carried out by the researcher from the last decades, and still now it is considered one of the most vital fields of research in the power sector. According to the researcher, it is claimed that with the innovation of a smart grid, effective and efficient power control is possible. The smart grid offers the facility of communicating load demand and their characteristic with the generation grid. In addition, the grid can predict the disturbance or even fault by time to time analysis and take necessary decisions by itself and also blow alarm in case of any emergency. Therefore, the continuity of supply is maintained. The smart grid can fulfill the demand of 21st century power generation by providing reliability, efficiency, economy, and other crucial parameters besides optimally utilizing the available power resources.

The rest of the chapter is organized as follows. The journey from the traditional grid to the smart grid is discussed in Section 2. Section 3 highlights the benefits of smart grid system. Section 4 describes the standards and technologies for smart grids. The implementation aspects are presented in Section 5. The issues and challenges in implementing a smart grid are highlighted in Section 6. The open research questions and future research directions are discussed in Section 7 and a brief conclusion in Section 8 ends the chapter.

## 2.2  EVOLUTION: FROM POWER GRID TO SMART GRID

In the early stage, power was generated by DC source and grids were isolated from the load, therefore DC power is transmitted over a short distance only because of higher power loss in the transmission line. At the end of the 19th century, with the advancement of alternating current (AC), it becomes possible to transmit power over a long-distance using transformers technology where voltages can be stepping up reduced power loss.

In the traditional power plant, electricity is generated from coal, oil, or petroleum gas utilizing the mediator of steam. All the generators are generally synchronous generators, having a lower number of the pole (two or four) and running at high speeds (1500-3600 rpm). The overall electrical energy conversion efficiency is greatly affected due to the poor efficiency of the turbine and condenser (achieved efficiency ranges from 30% to 40%). The main advantage of the conventional plants is the low

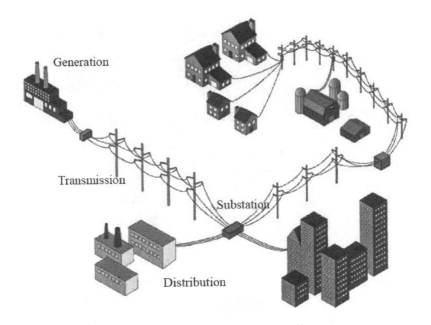

**Figure 2.3**  Traditional power system design

capital cost per kilowatt as compared to the other plants and there is no upper limit on their size [5].

The conventionally electrical power grid is designed to generate electrical power from a fixed location and then transmit the generated power through a high voltage transmission line and finally, the voltage is stepped down before supplying to the consumers. Figure 2.3 shows a typical diagram of the traditional power grid design. It transmits the generated power over a long distance to reach the consumers. The system consists of generators, transformers, the interconnection of transmission lines, distribution lines, substations, and consumers. Therefore, the traditional power system has three main sections.

**Phase 1 Power Generation:** This is an initial phase of the power system that requires high investment costs. The traditional grid has less control over the transmission grid and distribution feeders and also generation units have less control during peak hours and off-peak hours. Therefore, real-time monitoring for power generation becomes challenging [31]. Depending upon the distance of power generation, it can be divided into two categories, one is centralized and the other is decentralized. Centralized generation means power is generated far away from the load and consists of nuclear, coal, natural gas, hydro, large solar power plants whereas, if the power is generated near the consumer end e.g. power from mini rooftop solar plants is termed as a decentralized generation.

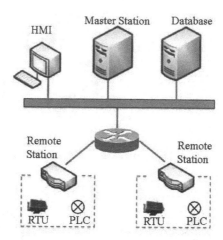

**Figure 2.4**   An example of a SCADA-based control system

**Phase 2 Power Transmission:** After power generation, it is transmitted for long distances to reach the distribution grid. It accounts for about 10% to 20% of the total cost of the electric power system. Hence, a significant amount of power loss occurs during the long transmission of power [31]. The transmission line consists of transformers and substations that are needed to transmit power from a generation plant to the consumer end. Transformers are used to step up the voltage initially to reduce power loss and then transmitting to long-distance through overhead or underground transmission lines. When this power reaches the distribution end, another substation is required to step down the voltage to the consumer rated voltage to deliver the power.

**Phase 3 Power Distribution:** In the distribution grid, power interruption and uncertainty is a common problem for the traditional grid as the distribution grid has almost no communication with the grid, hence consumers cannot understand the peak and off-peak hours. Electric bill is manually made by lineman and no real-time monitoring is possible [31]. Consumers can be categorized in many types depending on their rated voltage and power level. Distribution transformers are used to meet up the required voltage level for different types of load such as industrial, agricultural, residential and commercial consumers.

To meet up the load change or other parameter variation in the distribution grid, engineering calculations are required and the overall system needs to be recalibrated by changing the inputs of the model. However, electricity is generated in such a way that every time it matches the demand, therefore, to meet up the unexpected peak load some redundant power generators are used in parallel with the existing plants. Additionally, power is also stored in the battery when the demand is less and can be used to meet the excessive load. The battery technology is expensive and cannot able to meet up in a wider range of load variations. Therefore, the grid suffers from instability and power shortage. The power grid system is complex and interconnected with one another, hence, failure in one section can lead the overall system down [19].

In 1960, Supervisory Control And Data Acquisition (SCADA) framework, which appeared in Figure 2.4, is utilized to improve reliability and make the distribution framework partially automated. In spite of the fact that SCADA is not a particular innovation yet but is utilized for the modern procedure. The essential design of the SCADA framework is given in Figure 2.4 and basic components are discussed below.

> Remote telemetry units (RTUs): The system is liable for interacting with sensors and it collects the sensed data into standard form and finally sends data to the monitoring station.
> Programmable logic controller (PLCs): This device is used for control and automation, for example, the operation of circuit breakers are maintained by PLC.
> Master station: It is utilized to gather information from RTUs and examination it at that point sends commands to the PLCs.
> Database: It is used to store data for further analysis and to make a prediction in the future.
> Human Machine Interfaces (HMIs): After analyzing the data, it makes a simplified model to easily understand for the operator, hence, it helps to make a supervisory decision. A communication infrastructure connects the HMI with other SCADA components.

This SCADA system was installed in every substation and transmission network to connect the remote station with the grid [4]. With the updating technology, most vendors updated and added more features with the SCADA and provided the opportunity to better control and protect the power grid, for example, Modbus [44], IEC 61850 [12]. The architecture of SCADA was developed overtime to interconnect more SCADA through a wide area network [40]. However, the system is not suitable for implementing the fully automated next-generation power system.

In the traditional power system, it provides less information to the utility grid and has a few control from the generating terminal. Here, the transmission line is carefully monitored and a real-time monitoring system is not occurring, however, only some metering system is limitedly applied to the distribution grids to observe the supplied voltage and other parameters. Some generating stations can control some parameters like voltage and frequency remotely, however, most of the utilities use some remote equipment that can only sense voltage and other parameters and generate power from the predefined value.

Due to limited metering and communications from a substation and feeder point, it provides less information on line loss, voltage sag, transformer loading to the generating station. Additionally, traditional grid face challenge to integrate several renewable energy sources such as solar energy, wind energy, etc. due to lack of communication and remote control from the utility grid. The increasing number of electric vehicles also limits the capacity of the traditional grid.

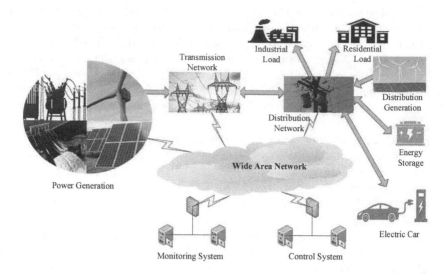

**Figure 2.5**   Illustrating a smart grid system

The traditional grid uses electromechanical meters that are operated manually, the electric bill is collected by the lineman once in a month and service holders get the bill with little detail or even the rate of the electric bill varies in peak time, therefore, many consumers including household and commercial face problems and also claim about the money of the bill. Hence, the smart grid system is essential to mitigate all these problems.

With the advancement of information technology (IT) and communication technology, it is also possible to apply this progressive knowledge into the power grid to make the system smarter. Additionally, the internet of things (IoT) also is used with the power grid to convert it into the modern generation grid. In this smart grid, there is a wide range of operational flexibility including generation, transmission, and distribution sections, and a user-friendly environment is created for both consumers as well as suppliers. Control can be done easily through IoT and computers.

Figure 2.5 shows the schematic diagram of a smart grid. The smart grid is a cyber-physical framework outfitted with achievable models of energy generation, distribution, and usage. The cyber-physical system is an integral component of the Internet of Things (IoT). The cyber part is the Internet (wide area network) which is intertwined very tightly with the physical system in order to make it smart so that it can do anything in real-time more efficiently than the traditional physical system.

The smart grid is concept that utilizes data innovation to convey power productively, dependably, and safely [33, 11]. The smart grid also known as:

Electricity with brain
The energy internet
The electronet

## 2.3   BENEFITS OF SMART GRID SYSTEM

Traditional power grids face a lot of challenges in managing power system especially in the generation, transmission, and distribution. Most of the problems of the traditional grid are possible to solve by the concept of the smart grid [49, 28]. The smart grid offers benefits to both customers and stakeholders for efficient operation and management of electricity. Additionally, it increases more reliability and continuity of supply by connecting the whole system in communication mode [8].

### 2.3.1   TECHNOLOGICAL BENEFITS

The technological benefits related to smart grid innovation are given as follow:

*Efficient power transmission:* The main advantage of the smart grid is that it ensures efficient power transmission over the long transmission line. Voltage parameters can be easily measured remotely by using different sensors and proper actions can also be taken to ensure the quality of power.

*Quicker restoration:* In case of any disturbance, the smart grid is capable of quick restoration compared to the conventional grid. Additionally, several minor faults can also be cleared automatically and maintain stability with the help of an Artificial Intelligence (AI) automated sensor system.

*Reduced cost:* Smart grid offers lower operation and the executive costs for utilities as well as lower power costs for consumers. In the smart grid, the devices are connected with each other in a cloud-based environment. Therefore, the task like follow up procedure and general data collection become easy and it will significantly reduce operational and maintenance cost.

*Reduced peak demand:* The smart grid provides a wide range of opportunities to communicate with smart devices including loads and generators in real-time. In peak time when the electric demands increase gradually, the user plays a vital role to reduce the peak demand to avoid higher electric tariffs which likewise helps to reduce peak power demands [30].

*Increased use of renewable sources:* The smart grid is an integrated platform that generates power from different sources including renewables. Therefore, depending upon the load conditions, power plants can extract maximum power from the renewables that significantly increase the huge scope to utilize the sustainable power source frameworks.

*Good integration of customers and energy providers:* A good bond between the consumers and energy providers are building up as the customers are directly involved with the smart grid and ensure both way communication frameworks.

*Improved security:* Smart grid provides improved security in terms of reliable and stable operation of power plant with real-time monitoring facilities. A multi-layer protection scheme is developed to guarantee security for participating consumers and suppliers [26].

## 2.3.2  BENEFITS TO CUSTOMERS

The smart grid offers many benefits to consumers, including the following:

*Real-time update on energy usage:* The customers can monitor their individual energy consumption on a continuous and real-time basis. Therefore, the customers also limit their high energy consumption to avoid the high charge of the electric bill. Additionally, consumers can easily predict their monthly electric bill, and a clean environment is created between the suppliers and consumers.

*Various pricing models:* The various pricing models and different tariff are characteristics of the smart grid. Therefore, it encourages the customers to use their heavy equipment's operation during off-peak hours to minimize the electric bill. During the peak-hours, a high electric charge is imposed to discourage consumers from high consumption of energy.

*Proper utilization of off-peak hours:* In the smart grid, smart devices such as electric cars can be charged during off-peak hours to enjoy reduced electric charge [27]. Consequently, that helps the easy grid operation and limits the peak demands of the plant.

*Energy contributors:* The smart grid is a two-way communication that provides a good chance for their customers to become an energy contributor. For example, customers can produce energy at their roof-top by installing PV-plants or small wind turbines that can also couple with the main grid. Hence, customers can be able to provide energy into the main grid and get financial benefits.

## 2.3.3  BENEFITS TO STAKEHOLDERS

The smart grid system offers several benefits to stakeholders, including the following:

*Increased grid reliability:* The smart grid offers more reliable and maintains continuity in grid operation compared to the conventional grid. The use of sensors in different devices increases better communication facilities and certainly, it accelerates the grid stability and reliability.

*Reduce the frequency of power outages and brownouts:* Load-shedding is becoming a common problem for many areas. One of the vital reasons for being load-shading is an unstable operation of the grid. In this situation, the smart grid provides a wider range of control to maintain the stable and safe operation.

*Provide good infrastructure:* Good operating infrastructure is another key feature of the smart grid. The smart grid framework provides good structural facilities for continuous supervision, examination, and self decision making in case of any minor parameter variations.

*Increase grid versatility:* As the sensors are employed in the smart grid in most of the equipment, it increases grid versatility by providing detailed data. The data is stored on a real-time basis in the cloud-based environment and analysis becomes easy to make any changes.

*Reduce inefficiency in power supply:* Highly efficient power transmission and distribution are possible in the smart grid. The smart grid offers an automated optimization technique that ensures efficient power transmission.

*Integrate renewable sources:* The smart grid not only depends on fossil fuels to generate power but also integrate the sustainable resources of wind and solar alongside the main grid. Therefore, it certainly cut down the high cost of power generation.

*Improve the management:* Improved working frameworks are possible in the smart grid system. The uses of distributed energy resources, including micro-grid operations and storage management make the grid more reliable and stable with smooth management.

## 2.4 SMART GRID: STANDARDS AND TECHNOLOGIES

### 2.4.1 STANDARDS

Many researchers have reported the importance of having standards for smart grid. However, one of the main challenges to designing a smart grid is the lack of standards that prevents the integration of advanced applications, smart meters, and smart devices [9]. It is important to consider the safety and security aspect of smart grid in addition to communication facilities and protocols [23, 35]. Many organizations work on developing the standards, the name of some standards bodies include European Union Technology Platform, Ontario Energy Board, Canada, NIST, the

American National Standards Institute (ANSI), ), the International Electro Technical Commission (IEC), the Institute of Electrical and Electronics Engineers (IEEE), the International Organization for Standardization (ISO), the International Telecommunication Union (ITU), the Third Generation Partnership Project (3 GPP), and so on. The overview of the smart grid standards are summarized in Table 2.1 [23, 29].

The details are discussed below:

**Table 2.1**

**Overview of Smart Grid Standards**

| Standards | Application |
|---|---|
| IEC 61970; IEC 61969 | Energy management system |
| IEC 61850 | Substation automation |
| IEC 60870-6/ TASE.2 | Inter-control center communications |
| IEC 62351 Parts 1-8 | Information security system |
| IEEE P2030 | Customer-side application |
| IEEE P1901 | In-home multimedia, utility, smart grid |
| ITU-T-G.9955; G.9956 | Distribution automation and AMI |
| Open ADR | Price responsive and load control |
| BACnet | Building automation |
| HomePlug | HAN |
| HomePlug; GreenPHY | HAN |
| U-SNAP | HAN |
| ISA 100.11a | Industrial automation |
| SAE J2293 | Electrical vehicle equipment |
| ANSI C12.22 | AMI |
| ANSI C12.18 | AMI |
| ANSI C12.19 | AMI |
| Z-Wave | HAN |
| M-Bus | AMI |
| PRIME | AMI |
| G3-PLC | AMI |
| SAE J2836 | Electrical vehicle |
| SAE J2847 | Electrical vehicle |

### 2.4.1.1   Revenue Metering Information Model

ANSI C12.19: ANSI standards ANSI C12.19 used for transmitting data between the device and computer in a binary mode of transmission by means of binary codes and XML content. This is used in the utility industry.

M-Bus: The European standard M-Bus is used to take the data of all different kinds of meter remotely. A general master database is connected with the utility meters that can be able to take reading periodically with the help of M-Bus.

ANSI C12.18: This is also an ANSI standard that is especially considered to provide both way communication facilities to the smart meters with the utility grid.

### 2.4.1.2   Building Automation

BACnet: BACnet is designed by the American Society of Heating, Refrigerating, and Air-Conditioning Engineers (ASHRAE) which is a communication protocol used in smart homes to make the home fully automated and computer-based control of all electrical loads within the house.

### 2.4.1.3   Substation Automation

IEC 61850: This is an open standard flexible device that works for maintaining the communication among the transmission grid, distribution grid, and the substation automation system. The features of the device provide the opportunity for improved monitoring, control, and protection of devices [41] which is compatible with the Common Information Model (CIM).

### 2.4.1.4   Powerline Networking

HomePlug: This is a powerline skill used to connect the smart devices with the HAN. This is also aimed at low-cost uses and it is considered a reliable connection of HAN between devices and smart meter.

HomePlug Green PHY: These specifications are launch as a cost-reduced power line network standard for smart grid applications used in residential area networking. This technology is developed by the Smart Energy Technical Working Group within the HomePlug Powerline Alliance.

PRIME: This is a worldwide powerline standard that offers multivendor interoperability and welcomes several entities to its body. Some companies have good experience in PLC technology and smart metering such as Advanced Digital Design, CURRENT Group, Landis+Gyr, STMicroelectronics, uSyscom, and ZIV Medida.

G3-PLC: This technology is launched by ERDF and Maxim, also a powerline communication, aims to secure robustness and reduced cost, cybersecurity to the smart grid in worldwide implementations.

## 2.4.1.5 Home Area Network Device Communication Measurement and Control

U-SNAP: Utility Smart Network Access Port (U-SNAP) used to provide communication protocols to connect HAN with the smart meters. U-SNAP fundamentally allows the standardization of a connector and serial interface and identifies the hardware interface.

IEEE P1901: This was developed by the IEEE sponsorship, can offer high-speed power line communication to meet the basic requirements of home theater, utility, and smart grid applications [21].

Z-Wave: This technology was developed by the Z-Wave Alliance provides an alternative solution for ZigBee. This becomes the equipment of interest for home automation due to its simple model, and low-cost feature.

## 2.4.1.6 Application-Level Energy Management Systems

IEC 61970 and IEC 61968: These two defined standards called Common Information Model (CIM), used for exchanging data between devices and networks whereas the former works in the transmission domain and the latter for distribution domain.

OpenADR: OpenADR was developed at Lawrence Berkeley Labs, a U.S. government research laboratory, used for the improvement of smart grid dynamic pricing, demand response, and grid reliability.

## 2.4.1.7 Inter-Control and Interoperability Center Communications

IEEE P2030: The IEEE standard board has developed the standard. This is a guideline for smart grid power technology and communication technology with the electric power system (EPS) and customer-side uses. The system is responsible for both way communication for reliable power generation, power delivery.

ANSI C12.22: This standard defines secure and confidential communication to transport ANSI C12.19 table data to the smart meters using AES encryption.

ISA100.11a: This open standard was designed by the International Society of Automation (ISA) mainly focuses on security, robustness, and management of the wireless network which is simple to use.

ITU-T G.9955 and G.9956: This two standard is developed for providing supports in outdoor and indoor communications over low or medium voltage line. The first one is defined for physical layer specifications and the second one is defined for data link specifications.

### 2.4.1.8  Cyber Security

IEC 62351: Security of the smart grid is considered one of the major challenging issues because the utilities are porn to get attack by cyber threats as it is bidirectional communication. IEC 62351 defines the cybersecurity of the smart grid.

### 2.4.1.9  Electric Vehicles

SAE J2293: This standard provides information about electrical vehicles and electrical vehicles supply equipment, this was established by the Hybrid Committee that is a part of SAE international.

SAE J2836: SAE J2836 used for the communication for plug-in electric vehicles and energy transfer between the grids.

SAE J2847: This maintains communication messages between PEVs and grid components.

## 2.4.2  TECHNOLOGIES

The smart grid is the collaboration and integration of several technologies to provide better operation in control and power management. Among these the main three technologies that involved in smart grid are highlighted below:

### 2.4.2.1  Storage Systems

Energy storage technology is the vital technology in the smart grid and numerous advancements have been developed to store energy with high effectiveness and long life cycle. This storage system increases the reliability of the continuity of load supply by enabling improved penetration of wind and solar power in the system. Some of the names of energy storage systems are given below [16]:

Electrochemical storages (as Lead-Acid; Zinc Bromine, ZnBr; Sodium-Sulfur, NaS; Vanadium Redox, VRB; Nickel-Cadmium, Ni-Cd; Li-ion batteries; Hydrogen Energy Storage, HES)

Electromechanical storages (as flywheels, Pumped Hydro, and Compressed Air Energy Storage, CAES)

Electrostatic storages (as ultra-capacitors, UC)

Electromagnetic storages (as superconducting magnetic energy storage: SMES)

There is a significant quantity of energy loss occurs due to the energy conversion and to store it in a storage device, the power loss also depends on its efficiency and flexibility of the system.

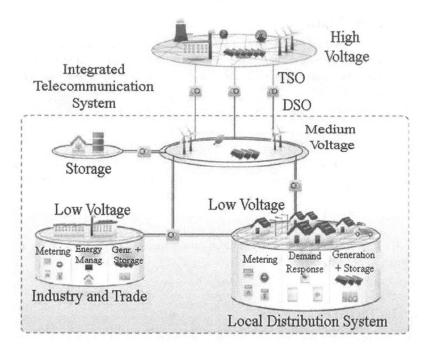

**Figure 2.6**   Smart grid integrated telecommunication systems

### 2.4.2.2   Telecommunication Systems

As mentioned in the 'Evaluation Section' traditional grid used SCADA communication which is generally based on direct serial, radio links, dial-up, or modern connections. However, the fundamental issue of SCADA is the lack of architectural facilities to handle multiple channels and a huge range of data. Therefore, to handle big data, Quality of Service (QoS) is required which is not implemented yet. To facilitate the greater flexibility, a smart grid internet-based network is needed but with some differences. There are two types of the network must be developed one is backbone network and the other one is the local-area network.

Infrastructural nodes are available in the backbone network whereas wireless connection is used in a local-area network provides a better opportunity for the smart grid to communicate with each other [39]. This system offers easy access to information, lower cost, less complication, flexibility, and the availability of off-the shelf wireless products such as Wi-Fi and ZigBee. This is an embedded product [48] and the huge number of traffic flow can easily be monitored [42]. Some vendors providing a solution that uses the internet that will guarantee them a reserved band or priority for their data transmission. In this way it is possible to have coverage of several physical layers and data links, mentioning to ISO/OSI protocol. Figure 2.6 shows how the grid can be split into three macro areas: low voltage, medium voltage, and high voltage.

Different technologies are different from one another and certainly be affected by a new combined telecommunication platform, for example, PLC is suitable for low voltage context such as to control the load in-home appliance. Now, many of the updated devices are available in the market, and most PLC smart grid applications on the low voltage side area of DSM, vehicle-to-grid communications, home energy management, etc. [20]

### 2.4.2.3 ICT Infrastructure

The information and communication technology helps to operate the power networking system incorporation with the SCADA model. The system contains the following[17]:

a Human-Machine Interface
a supervisory SCADA Master server
some remote units of the programmable logic controller
some intelligent electronic devices
the communications infrastructure that provides communications between the different device

The new SCADA system must be decentralized, profitable, in real-time, and updated. The real-time management viewpoint allows you to change network settings to improve overall network reliability and minimize the amount of damage where the load is removed to reduce the load on critical components.

In fact, the real-time presence of this last point and the location of an error to recover the service as soon as possible and hence:

Configure the MV network to reduce the number of disconnected clients
Send the staff to solve the problem

For the SCADA system, information security is considered a significant part of preventing cyber-attacks that could compromise a single stakeholder or the overall network. Therefore, a trusted smart grid requires a multi-layer security system with a cyber-infrastructure that bounds unfavorable entree and stable energy applications that can adapt to attack [45]. This system can benefit from other progress, such as adaptive protection and protection of the entire system.

## 2.5 IMPLEMENTATION ASPECTS

There are eight diverse progressive step for the fruitful execution of the smart grid and in this procedure consumers likewise assume an indispensable job appeared in Figure 2.7. The barriers in one step may lead to the cancellation of the implementation process of the smart grid [1]. The description of each step is given as follows:

**Step 1. Comprehend the drivers for smart grid implementation:** The customer point of view: As the consumers are habituated with the existing traditional grid, they

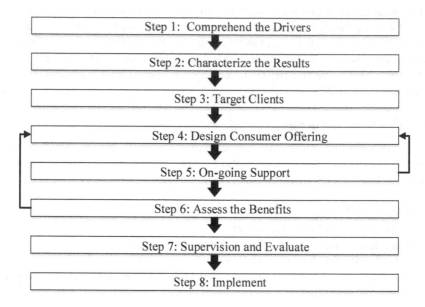

**Figure 2.7**  Implementation aspects of the smart grid system

may not feel comfortable moving on the smart grid. However, considering the aspects of consumers, it is not their responsibility to think about how the power companies provide sustainable, economic, and secure power supply rather than they can only consider the continuity of supply. Therefore, it is not sufficient to motivate the customers to change into the smart grid rather than conveying the customer's benefits if they move on. Additionally, it is also needed to understand what the customers expect and what they ask to do so.

> The context: Another important driver for implementing the smart grid is considering the location and the power context as it widely varies from place to place. Therefore, reference should be taken initially to the specific context before implementing the smart grid. This also includes the calculation of the total power need in the country not considering a precise zone. If not, there is a probability to increase conflicts that could lead to undermining overall effectiveness.
> Conflicts between stakeholders: This is true that the need for every customer may not be the same and they must not be aligned in the same directions. Therefore, it is important to identify and manage the needs of customers. Otherwise, it can lead to an obstacle to the expansion of the smart grid.

**Step 2. Characterize the results that consumers are required to convey:** This step defines the behavior change of the consumer that involves the initiatives of the smart grid. The expected outcomes are listed below:

Reduction in power consumption
Different patterns of power consumption may arise
Consumers' demand profile may vary from time to time and depending on
seasons
By using the smart meters information sharing method between the grid and
customers may provide efficient energy consumption

Initially, many of the readers think that it may also happen with the conventional
grid, however, these are not possible without the direct involvement of consumers in
the smart grid. Consumers can fix their tariff and it allows space for them to think
about whether they change their energy consumption behavior or not.

**Step 3. Target clients who could convey the necessary results:** It is also significant to identify the target customers of the smart grid and it is expected that:

Some clients are more ready to consider the interest reaction than others
Some clients are more tolerating of Demand Side Response than others

Figure 2.8 shows the process flow chart of the smart grid for step 3. After com-
pleting steps 1 and 2 it passes through some decision boxes where several questions
are analyzed and decisions are taken upon the answer to the questions. Finally, the
system reaches step 4 through "Target coincident load selection" and "Customer
Segmentation".

**Step 4. Design customer offering:** How the customers can willingly engage with
the project to speed up the implementation process of the smart grid is going to
discuss in this step. Some of the important concern for designing customer offering
is given below:

Ensure initiative provides tangible benefits: Several studies have been performed and also on-going to determine the benefits of implementing a smart
grid for both customers as well as power companies. It is important to ensure
specific benefits for consumers.
Leverage key events: Numbers of motivational events are organized to influence the customers to sign-up with the smart grid projects. Two such cases
are:

**- when clients are thinking about exchanging their energy provider:**
It is observed that the rate of customers willing to come forward to change the energy pattern by themself is rare, therefore, for the power companies, it is essential to
inform their skate holder about new tariffs or recruiting new customers with some additional offers. This scheme also increases consumers' awareness of effective power
consumption.

**- when clients are purchasing new appliances:**
When the customers go to electric showrooms to buy some new pieces of equipment like washing machines or televisions, they can learn about the new power
scheme and its' benefits from the seller.

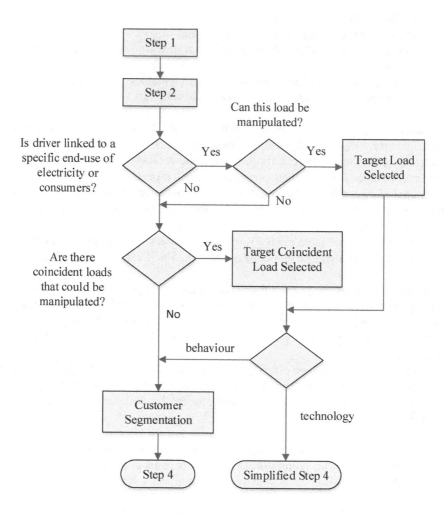

**Figure 2.8** Process flow chart of step 3

Can the activity answer a current issue?: At first, this needs to be figured out whether the smart grid can explain the current difficulty of the traditional grid, and this message needs to convey to the customers.

Community engagement: Community meetings can be an effective way to engage and motivate more customers in a short time. This also helps power companies to get funds or in-direct benefits from the customers.

Provide access to reliable and trustworthy information: Good communication between stakeholders and utilities is important. Hence, it will increase faith and reliability.

Provide choice: It is a wise thing to provide a choice for the customers when smart grid initiatives are taken into considerations. Otherwise, the failure of a small part would raise a protest against the smart grid.

**Step 5. On-going support:** This is a continuous feedback process and initiatives taken form the utility part to confirm the customers to signed-up with the execution system of the smart grid. This helps to increase the success rate to implement a smart grid. This approach includes:

Guaranteeing any new customer concerns are distinguished as they emerge
Giving on-going help to guarantee results are conveyed

**Step 6. Assess the benefits:** The smart grid can be capable to provide advantages for both the energy system and overall society.

In any case, the target of this activity isn't to measure the general advantages of the smart grid, however, to survey the advantages from the clients' point of view of the smart grid activities engaged with the modified system. Along these lines, this segment considers the accompanying issues:

The smart grid identifies the benefits that companies offer customers
Delivery of benefits derived from changes in customer behavior

But before doing so, it is important to first consider whether it needs to directly benefit the customer or whether it is enough to benefit society as a whole.

**Step 7. Supervision and evaluation:** It is considered significant to monitor and evaluate the benefits for customers in and every stage to get more motivated customers. From the utilities, it is not only important to measure the energy generated cost to describe the project is successful or not, however, assess what types of initiative is proved more successful.

The surveys and study on the customers are analyzed and how customers are connected in the project need to justify. How the customers react with this new technology, their feedback on new tariffs also be noticed for further development of the grid.

The after effect of a smart grid activity (for example the result characterized in step 2) will rely upon the conduct of customers. This, thus, is driven by a few components identifying with the individual and the cultural setting where they work.

**Step 8. Implementation:** The step-by-step approach outlined here is believed to represent the components that should be routed to guarantee that customers are most interested in "subscribing" to smart grid activities and delivering the expected results. Each progression tends to various parts of the energy conduct model. Specifically, it guarantees that the savvy matrix activity is intended to:

> Recognizable advantages are conveyed to customers
> Explicit requirements of the important business are met
> The outcomes are observed to survey which parts have been fruitful, why they have been effective and for whom they have been fruitful

Finally, with the active participation of the customers, it is expected that the smart grid project will show the path of success.

## 2.6    CHALLENGES OF IMPLEMENTING SMART GRID

To design a smart grid is not free from challenges like other technologies as it includes technical, social, and economic constraints. Technical challenges are generally considered to integrate several distributed generators including renewable and non-renewable with the grid where social acceptance is also becoming an important issue. Sometimes money/funds are considered the most important barrier to design a smart grid. The details of all three challenges are given below:

### 2.6.1    TECHNICAL CHALLENGES

> Inadequacies in framework foundation: In most creating nations, the network framework isn't sufficiently adequate to oblige the upcoming needs of the smart grid [32]. Several challenges may arise in the design, erection, operation, and maintenance. Some grids are not possible to convert from the old grid to the modern smart grid as there is less opportunity to do so due to space facilities. The re-installation of the plant is somehow mandatory and proper relocation of the heavy types of equipment needs healthy cost and time.
> Cyber security: To connect the grid into a cyber network caused the number of challenges in the system spacially due to security and communication issue. In the research paper [32], the authors mainly addressed and discussed three major issues, i.e. availability, integrity, and confidentiality. Availability implies dependable and ideal access to the database and other data. Integrity alludes to insurance from inappropriate organization of data, though, confidentiality incorporates giving security from unapproved access to the data. Cyber-security is viewed as one of the huge issues for the grid operation as if any hack has happened is becomes a matter of disaster for the utilities as well as for the consumers. There are some common cyber threats are hackers, zero-day, malware, etc. These challenges are not easy to handle if they are applied to the grid. A smart grid consists of a multilayer of structure and

each layer is defined for specific security concerns. The cyber threats are not limited to a certain type of number therefore, it is required to build-up higher technology to provide good protection against these threats [36].

Storage concerns: The smart grid is incorporated with renewable energy sources and as renewable energy is not uniform i.e. its output power as well as voltage or frequency has a varying characteristic, it needs to store in a storage device like battery, however, the life cycle of battery is limited maximum 4-5 years. There are some other storage technologies like flywheels, thermal storage, hydrogen storage, etc. are also available. Some country prefers to use pumped storage technique which has efficiency in the range of 70-85%. However, the main problem of the pumped storage method is that it requires a large area as reservoirs which are normally available only in the mountainside. Generating electricity from offshore and onshore are becoming popular in recent time as wind turbines are capable to extract a huge amount of energy from wind. However, to store this energy sometimes become challenging. To absorb and deliver energy quickly into the grid, many researchers suggested flywheel technology, however, it is not stable for a longer duration. Batteries are mostly used to store energy and it is considered the most common traditional method as batteries are portable, however, they suffer from low energy density, high weight, and size. Additionally, batteries need to maintain manually on a regular basis. To increase the efficiency and reduce the cost of the battery is mandatory because battery storage technologies are expensive. For a large scale of storage, Advanced Lead Acid Batteries, Flow Batteries, and Lithium-Ion Batteries are the other options but they are expensive also [43].

Data management: Real-time meters, sensors, and controllers are used to make the grid smart in collaboration with the information source such as security cameras, weather forecasts, etc. that can successively increase the capacity of the operators [6].

By the precise examination of the information, breakdown or damage could be maintained before occurrence. Also, this large information can be utilized for framework activity, alerts, forecasting demand, generation, cost, and so forth. The data is huge, for example, the meter that traditionally billed once in a month, the meter of the smart grid takes the reading after 15 minutes increased the number of data approximately 300 times. In the grid, although the huge amount of data is not too much problem to collect and store but to process the data it is becoming challenging, therefore, database supervision is a vibrant issue in SG. The overall process of the smart grid may slow down to handle and process large amounts of data and define the standers. In this scenario, cloud-based technology may be helpful to process big data.

Communication issues: Although, with the wide range of communication technologies available to the growth of the smart grid, each and every technology suffers from certain limitations. Some are suffering from limited bandwidth while others are suffering from higher data loss and lesser efficiency with the increasing distance. Therefore, to develop a smart grid, communication sectors still suffer lots of challenges even most of the communication rules are not very much characterized in the SG network. Some technologies like GSM, GPRS, PLCC, 3G, ZigBee, Broadband over PLC, etc. GSM and GPRS have a coverage range of up to 10 km but they lack data rates. 3G involves the costlier spectrum, while ZigBee has approximately 30 m to 50 m coverage range. To solve the distance problem wired communication system like power line communication can be used, however, it also faces a problem of interferences. Some other researchers proposed optical fiber communication but again this is expensive. Another researcher proposed RF technology with router-based, however, it also suffers from performance issues [23].

Stability concerns: Distributed generators (Renewable sources) and microgrids are the parts of the smart grid on a large scale of power plants. Distributed sources cause bidirectional power flow. In terms of green and low-cost energy renewables and microgrid have some certain advantages over the coal or nuclear power plant [22]. However, as it has the property of bidirectional power flow, with the penetration of a large power plant, it may cause stability problems such as:

Angular dependability because of low inertia
Voltage stability because of low power-sharing assistance
Low-frequency power wavering
Worsening of SG transients profile during MG islanding
Inability to fill in as framework hold

Energy management and electric vehicle: Nowadays, the number of electric vehicles is increasing and most of the user does not follow any restriction to charge its battery [37]. Smart grid influence the customer to charge it on an off-peak hour and can also be used as a source during peak hour [31]. That certainly reduces the electric bill for the customers and reduces the excessive presser on the grid. In this energy management, there is a certain lack to define proper standards including:

The flow of power from
– vehicle to grid (V2G),
– grid to vehicle (G2V),
– vehicle to vehicle (V2V).
Controlling of the reactive power
Controlling of the DC linked voltage
Supporting of the grid voltage

## 2.6.2  SOCIO-ECONOMIC CHALLENGES

To develop a new project, the socio-economic challenges become the most vital issue and it can lead to the failure of the pilot project and rejection of novel technologies due to lack of investors and less awareness among the consumers [24, 25]. The detail discussion of these challenges are given below:

High capital investment: Establishing a smart grid or converting the traditional grid into the smart grid requires a lot of initial investment. Although it can overcome with the long-term plan, it is difficult to find investors. For this situation, mindfulness projects and motivators are basic to energize utilities, associations, and people to comprehend the advantages of the smart grid. Power companies incorporation with the government and stakeholders can contribute proportionally to overcome this problem.

Stakeholder's engagement: Consumers can be discouraged with the new technology due to high capital investment and lack of information they know about the technology. Therefore, it is necessary to deliver accurate data to the customers and proper concealing is necessary to understand the positive sides of the smart grid.

System operation aspects: A smart grid also needs to define certain operational issues like billing, tariff structure, and operational strategies with its customers. Because customers mainly depend on transparency, policies, and reduced billing structure. Hence, from the part of system operational aspects, it is necessary to define the relevant terms and be flexible in every step in every step for the development of the smart grid.

Lack of awareness: As the smart grid is a new concept, many of the people raise their voices against it as they are happy with the traditional grid system. Hence, proper education is needed to raise customer awareness about the smart grid. The customers need to know the benefits of the smart grid over the traditional grid and how this method helps them to make reduce electric bills. Seminars and customer counseling can solve the problem partially to understand how the smart grid plays a vital role to boost up overall efficiency and economy. On the other hand, policymakers also need to consider the present and future scenario of the technology to introduce a flexible offer for the stakeholders.

Privacy: Privacy is important for both end, from the service provider and the consumers also. Smart grid often suffers from cyber threats and within a second huge data can be stolen and modification can also be done by the cyber-attack. Therefore, to maintain the faith of the customers, the utility grid needs to provide security against all kinds of threats.

Fear of obsolescence: Day by day, with the development of new technology, old products are gradually omitted, however, this is not like that the new product cost is low but customers are pressurized to buy new products at a higher price. It is also predicted that the development of IT and communication technology can be a barrier to the expansion of the smart grid.

Fear of electricity charge increase: As the smart grid needs a high amount of initial investment, the customers are assumed that they need to pay a higher tariff than pay now with the traditional grid. Even the customers are not familiar with the upcoming facilities of the smart grid. Therefore, from the part of stakeholders, the fear of high tariffs also becomes a burden to the smart grid.

New tariff: In the scheme of the smart grid, the variable tariff may be introduced, a low price in off-peak hours, and a high price for peak hours. A new tariff may not be accepted for all customers. Some customers consume electricity when needed without considering the time. On the contrary, in the new tariff, real-time monitoring of electricity is involved. This may be annoying for customers.

Radio-frequency (RF) sign and medical problems: Some medical specialists and health-conscious people claim that RF signal is transmitted from the smart grid devices, however, there is no exact data for analysis currently, it also can be a great issue and becomes an obstacle in the growth of the smart grid. Proper research and awareness initiatives are needed to handle this issue.

### 2.6.3  MISCELLANEOUS

Along with technical and socio-economic challenges, some other issues can be obstacles to introduce the smart grid into the society [6, 24]. Among the few important issues are discussed below:

Regulation and policies: Motivating stakeholders is very important for the newly implemented smart grid to take an active part in the development process which is mostly under research. The framework of the design is defined in such a way that we can maximize the contribution in each sector and the risk can be shared with the stakeholders according to their importance chain. In the guidelines and rules, we must study the following issue:

The implementation cost of smart grid must be shared among government, power companies, and stakeholders

Proper motivation is needed to encourage smart grid

Identify standards and roadmaps

Define target customers and achievable time frame

Meet up designations in each of the sectors including investors and consumers

Highly focus on awareness initiatives

Power theft: In most of the developing countries bypass line or power theft is becoming a common phenomenon because the metering system, billing technique, and charges collection are not well structured, therefore the AT&C loss of these countries is comparatively high approximately 10-12%. To handle this situation, technical and regulatory measures need to be defined and social initiative also recommended to motivate people against this unethical practice.

Workforce: As smart grid technology involves a wide range of expertise fields including power engineering, communication engineering, IT, and IoT, therefore, sometimes it becomes challenging to cooperate all these expertise in one field. More expert hand and training sessions are needed to handle this situation. Additionally, the manpower that is working in the traditional grid would be difficult to handle in the smart grid due to a lack of knowledge with the automation system.

Co-ordination: To handle a plant-like smart grid where major sections of engineering sectors are involved with non-technical people like policy-makers, manufacturers, business markets, economist, and also consumers sometimes becomes difficult for the proper co-ordination, therefore, misunderstand or misleading can cause a serious problem. Hence, proper coordination is needed among the different levels of people for the successful operation of the plant.

## 2.7   OPEN RESEARCH QUESTIONS

As the concept of smart grid is new, there are lots of wide opportunities for the researcher for its development [47, 18, 46, 15, 3]. The smart grid is the combination of power sectors, communication sectors, information technology (IT), and the internet of things (IoT). Therefore, to combine all these things in the same platform can be a big challenge. Some open research problems in smart grid design and implementation are highlighted below.

Smart grids are dealing with the real-time monitoring system, therefore, the number of data increases dramatically. To handle this big data is becoming a challenging task and sometimes the overall system slows down and fails to operate. Hence, research is needed in this area to overcome this issue.

New communication infrastructure is needed to develop to meet up the requirements and needs for self-healing grids. Development is required in this field to improve the reliability and power quality of the grid.

Power flow optimization is one of the challenging fields of research. To integrate all parameters together and to generate is precise output is mandatory for the improvement in power flow.

Renewable sources are interconnected with the smart grid and due to the non-linearity behavior of renewable energy, it is required to store the energy in EV battery and then connected with the grid. However, the batteries are expensive and the life cycle is minimum. Research is needed to prolong the battery life or what would be the replacement of the battery.

The concept of cloud-based power control and management is new and lots of issues need to handle and to manage a large number of data in the IoT server. Research is needed on how to handle the overall server effectively.

Security is one of the vital aspects of research for both the consumers and utilities. As all the information will be stored and handled from the cloud server it will suffer the threat of cyber-attack and with a minor imbalance, the whole grid may be affected seriously and data can be stolen and moderated that would create extreme harm for the system. Therefore, it is vital to develop a strong multi-layer protection scheme for the grid to handle different types of cyber attacks.

## 2.8 CONCLUDING REMARKS

In the 21st century, most of the grids try to convert traditional old age power grid into a smart grid for effective and efficient power control and to solve the challenges that are suffered by the traditional grid. However, conversion of the grid is not an easy task due to it requires lots of financial investments as well as research is also recommended before conversion. In this process, the grid needs to overcome certain challenges including technical and social problems that are also discussed in this chapter. Technical problems can be overcome by research and analysis by expertise hands however, to solve the social problem collaboration of the Governments with power companies as well as consumers are required to raise awareness among the people. Without understanding the benefits of the smart grid no one can be interested to invest resources to shift from the traditional grid to a smart grid. Therefore, proper counseling and seminars are also suggested to spread the knowledge of smart grid.

The expectation from the smart grid is getting wider with the progress of communication and information technology, overcoming all the obstacles, it is expected that the evaluation of smart grid in 21st century will be a milestone for a digitizing power system. With the automation, all things are getting connected with one another provides the opportunity to centralized control and effective management of the devices. Additionally, sooner the grid will be considered as a "grid of things" with the availability of internet as like "Internet of everything".

## REFERENCES

1. http://www.ieadsm.org/wp/files/smart-grid-implementation-version-1.0-issued-02-june-2014.pdf.
2. https://en.wikipedia.org/wiki/World\_energy\_consumption.
3. Syed Saqib Ali and Bong Jun Choi. State-of-the-art artificial intelligence techniques for distributed smart grids: A review. *Electronics*, 9(6):1030, 2020.
4. Ahlam Althobaiti, Anish Jindal, and Angelos K Marnerides. Scada-agnostic power modelling for distributed renewable energy sources. In *2020 IEEE 21st International Symposium on "A World of Wireless, Mobile and Multimedia Networks" (WoWMoM)*, pages 379–384. IEEE, 2020.
5. Emilio Ancillotti, Raffaele Bruno, and Marco Conti. The role of communication systems in smart grids: Architectures, technical solutions and research challenges. *Computer Communications*, 36(17–18):1665–1697, 2013.
6. George W Arnold. Challenges and opportunities in smart grid: A position article. *Proceedings of the IEEE*, 99(6):922–927, 2011.
7. Gagangeet Singh Aujla. Renewable energy based efficient framework for sustainability of data centres. 2018.
8. Gagangeet Singh Aujla, Sahil Garg, Shalini Batra, Neeraj Kumar, Ilsun You, and Vishal Sharma. Drops: A demand response optimization scheme in sdn-enabled smart energy ecosystem. *Information Sciences*, 476:453–473, 2019.

9. Gagangeet Singh Aujla, Anish Jindal, and Neeraj Kumar. Evaas: Electric vehicle-as-a-service for energy trading in sdn-enabled smart transportation system. *Computer Networks*, 143:247–262, 2018.

10. Gagangeet Singh Aujla, Neeraj Kumar, Mukesh Singh, and Albert Y Zomaya. Energy trading with dynamic pricing for electric vehicles in a smart city environment. *Journal of Parallel and Distributed Computing*, 127:169–183, 2019.

11. Gagangeet Singh Aujla, Mukesh Singh, Neeraj Kumar, and Albert Zomaya. Stackelberg game for energy-aware resource allocation to sustain data centers using res. *IEEE Transactions on Cloud Computing*, vol. 7, no. 4, pp. 1109–1123, 2017.

12. International Electrotechnical Commission et al. Iec 61850: Communication networks and systems in substations. *IEC*, 61850(2), 2004.

13. D Datta, MRI Sheikh, MM Islam, and A Al Mahdi. Terminal voltage regulation & output fluctuations minimization using governor-exciter controller. In *2016 2nd International Conference on Electrical, Computer & Telecommunication Engineering (ICECTE)*, pages 1–4. IEEE, 2016.

14. Dristi Datta, Md Rafiqul Islam Sheikh, Subrata Kumar Sarkar, and Sajal Kumar Das. Robust positive position feedback controller for voltage control of islanded microgrid. *Energy*, 1:3, 2018.

15. Illia Diahovchenko, Michal Kolcun, Zsolt Čonka, Volodymyr Savkiv, and Roman Mykhailyshyn. Progress and challenges in smart grids: Distributed generation, smart metering, energy storage and smart loads. *Iranian Journal of Science and Technology, Transactions of Electrical Engineering*, 1–15, 2020.

16. Francisco Díaz-González, Andreas Sumper, Oriol Gomis-Bellmunt, and Roberto Villafáfila-Robles. A review of energy storage technologies for wind power applications. *Renewable and Sustainable Energy Reviews*, 16(4):2154–2171, 2012.

17. Maria Carmen Falvo, Luigi Martirano, Danilo Sbordone, and Enrico Bocci. Technologies for smart grids: A brief review. In *2013 12th International Conference on Environment and Electrical Engineering*, pages 369–375. IEEE, 2013.

18. Mina Farmanbar, Kiyan Parham, Øystein Arild, and Chunming Rong. A widespread review of smart grids towards smart cities. *Energies*, 12(23):4484, 2019.

19. RG Farmer and EH Allen. Power system dynamic performance advancement from history of North American blackouts. In *2006 IEEE PES Power Systems Conference and Exposition*, pages 293–300. IEEE, 2006.

20. Stefano Galli, Anna Scaglione, and Zhifang Wang. For the grid and through the grid: The role of power line communications in the smart grid. *Proceedings of the IEEE*, 99(6):998–1027, 2011.

21. Shmuel Goldfisher and Shinji Tanabe. IEEE 1901 access system: An overview of its uniqueness and motivation. *IEEE Communications Magazine*, 48(10):150–157, 2010.

22. Pathirikkat Gopakumar, M Jaya Bharata Reddy, and Dusmanta Kumar Mohanta. Letter to the editor: Stability concerns in smart grid with emerging renewable energy technologies. *Electric Power Components and Systems*, 42(3–4):418–425, 2014.

23. Vehbi C Gungor, Dilan Sahin, Taskin Kocak, Salih Ergut, Concettina Buccella, Carlo Cecati, and Gerhard P Hancke. Smart grid technologies: Communication technologies and standards. *IEEE Transactions on Industrial Informatics*, 7(4):529–539, 2011.

24. Christopher Guo, Craig A Bond, and Anu Narayanan. *The adoption of new smart-grid technologies: Incentives, outcomes, and opportunities*. Rand Corporation, 2015.

25. Anu Gupta. Consumer adoption challenges to the smart grid. *Journal of Service Science (JSS)*, 5(2):79–86, 2012.

26. Anish Jindal, Gagangeet Singh Aujla, Neeraj Kumar, and Massimo Villari. Guardian: Blockchain-based secure demand response management in smart grid system. *IEEE Transactions on Services Computing*, 13(4):613–624, 2020.

27. Anish Jindal, Neeraj Kumar, and Mukesh Singh. Internet of energy-based demand response management scheme for smart homes and phevs using svm. *Future Generation Computer Systems*, 108:1058–1068, 2020.

28. Anish Jindal, Neeraj Kumar, and Mukesh Singh. A unified framework for big data acquisition, storage, and analytics for demand response management in smart cities. *Future Generation Computer Systems*, 108:921–934, 2020.

29. Anish Jindal, Angelos K Marnerides, Antonios Gouglidis, Andreas Mauthe, and David Hutchison. Communication standards for distributed renewable energy sources integration in future electricity distribution networks. In *ICASSP 2019–2019 IEEE International Conference on Acoustics, Speech and Signal Processing (ICASSP)*, pages 8390–8393. IEEE, 2019.

30. Anish Jindal, Mukesh Singh, and Neeraj Kumar. Consumption-aware data analytical demand response scheme for peak load reduction in smart grid. *IEEE Transactions on Industrial Electronics*, 65(11):8993–9004, 2018.

31. Pratik Kalkal and Vijay Kumar Garg. Transition from conventional to modern grids: Modern grid include microgrid and smartgrid. In *2017 4th International Conference on Signal Processing, Computing and Control (ISPCC)*, pages 223–228. IEEE, 2017.

32. Ramakrishna Kappagantu and S Arul Daniel. Challenges and issues of smart grid implementation: A case of Indian scenario. *Journal of Electrical Systems and Information Technology*, 5(3):453–467, 2018.

33. Devinder Kaur, Gagangeet Singh Aujla, Neeraj Kumar, Albert Y Zomaya, Charith Perera, and Rajiv Ranjan. Tensor-based big data management scheme for dimensionality reduction problem in smart grid systems: Sdn perspective. *IEEE Transactions on Knowledge and Data Engineering*, 30(10):1985–1998, 2018.

34. Benjamin Kroposki, Robert Lasseter, Toshifumi Ise, Satoshi Morozumi, Stavros Papathanassiou, and Nikos Hatziargyriou. Making microgrids work. *IEEE Power and Energy Magazine*, 6(3):40–53, 2008.

35. Neeraj Kumar, Gagangeet Singh Aujla, Ashok Kumar Das, and Mauro Conti. Eccauth: A secure authentication protocol for demand response management in a smart grid system. *IEEE Transactions on Industrial Informatics*, 15(12):6572–6582, 2019.

36. Rafał Leszczyna. Standards on cyber security assessment of smart grid. *International Journal of Critical Infrastructure Protection*, 22:70–89, 2018.

37. Shuhui Li, Ke Bao, Xingang Fu, and Huiying Zheng. Energy management and control of electric vehicle charging stations. *Electric Power Components and Systems*, 42(3–4):339–347, 2014.

38. Alvaro Llaria, Octavian Curea, Jaime Jiménez, and Haritza Camblong. Survey on microgrids: Unplanned islanding and related inverter control techniques. *Renewable Energy*, 36(8):2052–2061, 2011.

39. Salman Mohagheghi, J Stoupis, and Z Wang. Communication protocols and networks for power systems-current status and future trends. In *2009 IEEE/PES Power Systems Conference and Exposition*, pages 1–9. IEEE, 2009.

40. Daniel E Nordell. Communication systems for distribution automation. In *2008 IEEE/PES Transmission and Distribution Conference and Exposition*, pages 1–14. IEEE, 2008.

41. Thilo Sauter and Maksim Lobashov. End-to-end communication architecture for smart grids. *IEEE Transactions on Industrial Electronics*, 58(4):1218–1228, 2010.
42. Tarlochan S Sidhu and Yujie Yin. Modelling and simulation for performance evaluation of iec61850-based substation communication systems. *IEEE Transactions on Power Delivery*, 22(3):1482–1489, 2007.
43. B. Easton, and J. Byars Smart grid: A race worth winning? A report on the economic benefits of the smart grid, 2012.
44. Modbus Application Protocol Specification. V1. 1b. *Modbus Organization*, 2006.
45. Siddharth Sridhar, Adam Hahn, and Manimaran Govindarasu. Cyber–physical system security for the electric power grid. *Proceedings of the IEEE*, 100(1):210–224, 2011.
46. Hongbin Sun, Qinglai Guo, Junjian Qi, Venkataramana Ajjarapu, Richard Bravo, Joe Chow, Zhengshuo Li, Rohit Moghe, Ehsan Nasr-Azadani, Ujjwol Tamrakar, et al. Review of challenges and research opportunities for voltage control in smart grids. *IEEE Transactions on Power Systems*, 34(4):2790–2801, 2019.
47. Maria Lorena Tuballa and Michael Lochinvar Abundo. A review of the development of smart grid technologies. *Renewable and Sustainable Energy Reviews*, 59:710–725, 2016.
48. Qiang Zhang, Yugeng Sun, and Zhenhui Cui. Application and analysis of zigbee technology for smart grid. In *2010 International Conference on Computer and Information Application*, pages 171–174. IEEE, 2010.
49. Jinju Zhou, Lina He, Canbing Li, Yijia Cao, Xubin Liu, and Yinghui Geng. What's the difference between traditional power grid and smart grid?—from dispatching perspective. In *2013 IEEE PES Asia-Pacific Power and Energy Engineering Conference (APPEEC)*, pages 1–6. IEEE, 2013.

# 3 Smart Grids: An Integrated Perspective

*Rafael S. Salles*
Institute of Electrical and Energy Systems
Federal University of Itajubá

*B. Isaías Lima Fuly*
Institute of Electrical and Energy Systems
Federal University of Itajubá

*Paulo F. Ribeiro*
Institute of Electrical and Energy Systems
Federal University of Itajubá

## CONTENTS

## 3.1 INTRODUCTION

The Electrical Power Systems around the world are based on large-scale production and are mainly centralized. However, this is a changing reality. The systems are adapting to the growing spread of distributed generation. Thus, the bulk power generation is now complemented by a significant number of small scale power generation units and Microgrids. Another characteristic of this new generation context is the high penetration of renewable energy sources, which is explained by society's greater concern with Global Warming and the reduction of $CO_2$ emissions.

Along with these transformations, the systems that make up the energy supply chain have acquired new complexities in the operation, carrying some issues, such as the insecurity of renewable generation, new problems of supply quality, new paradigms of the electric sector, and new demands. Thus, to deal with these issues, more robust solutions are required, in addition to the use of new technologies that make the grid become smarter. The relentless evolution of communication systems, in particular, has derived the development of advanced control and monitoring techniques to increase the reliability and quality of energy systems, in addition to making it more financially advantageous.

Smart Grids (SG) proposes a structure with evolutionary features of the network to adapt to the new scenario with a wide use of information and communication technologies (ICTs), bidirectional flow of energy and data, use of artificial intelligence, among other aspects [1, 2]. This modern grid also promotes better integration between microgrids and large-scale systems. This will require a major change in the operational paradigm, evolving from an outdated passive perspective toward a future grid active management approach [3].

This transition incorporates the large insertion of renewable sources, with an emphasis on solar and wind energy in the context of distributed generation. This aspect has several advantages when it is considered in regards to the consumer, who in this scenario becomes a prosumer, that is, one who can consume as well as produce energy. Looking at it this way, this agent now has more freedom to manage its energy balance. Another benefit is related to the use of clean energy which is not harmful to the environment and capable of reducing the carbon footprints through the use of electricity. On the other hand, the intermittent characteristics of these sources and their high penetration leads to many technical concerns, such as rapid voltage variations, reactive power control, harmonic distortions, quality issues.

As a result of the various aspects mentioned, economic changes are also foreseen in the electricity sector, some of which already takes place to provide the benefits of SGs on time. Along with the technical challenges, new business models and new regulations are emerging to change the rules following the new scenario, without losing the previously established social welfare premises. Social and political considerations are discussed for new projects, since the affected stakeholders are quite varied. News in the consumer classes, transaction models, rate plans are examples that promote this adequacy, but that cannot fail to consider universal access to energy and the market peculiarities. The services provided change, for example, the reactive power control at distribution site to mitigate the effects of the massive use of photovoltaic (PV) inverters could be a new auxiliary service.

Several applications appear in this context to characterize the SGs. Given the various variables that influence the systems, the load management techniques are increasingly improved to reduce costs and increase efficiency [4]. Internet-based applications become common place, looking for more participation by network users and also the development of large databases. Among these applications and others, the use of Artificial Intelligence (AI) becomes very common at different levels of network interaction. These techniques allow us to improve the existing services and also deal with the most modern technologies. Machine Learning, for example, is a technique with great potential for the operation and measurement of smart devices, such as Smart Meters, which is a disruptive technology in this evolution.

Thus, the grids of the present are being transformed to become highly computed, with advanced applications and information technology to deal with the challenges of improving energy systems [5]. These characteristics make the power grid systems integrated to various applications in urban environments, forming part of so-called Smart Cities. Present common approaches on the use of technologies and applications focused on end-users, where information plays an important role through the large-scale use of communication systems. The proposal of these different levels of systems has objectives that aim to modernize spaces and the way people relate to each other, to create a more sustainable, intelligent, and safe society.

The interaction of these various systems, equipment, and structures, requires special attention for the expected functioning of the grid. The lack of interoperability between the various elements is often the main obstacle to expand and transferring solutions that have been proven in a given network, city or system [6, 7, 8].

Having presented this overview of what SGs are, it is important to highlight some practical details that contextualize the content, a list is shown below:

1. Greater efficiency of transmission and distribution systems;
2. Ability to meet growing demand without adding significant infrastructure;
3. Ability to integrating better ideas and technologies;
4. Lower peak demand and favor better electricity rates;
5. Integration of customer-owner power generation units;
6. Focus on quality to eliminate disturbances in the grid;
7. Strong against attacks and natural disasters.

These listed features are in line with the Smart Grid idea discussed. These points show that the new network model prioritizes energy efficiency and intelligent customer service, where there are incentives and technological adequacy for the new needs. Also, in summary, the focus on quality and reliability guarantee starts to increase so that operations are leaner even during unforeseen problems. Another point is the capacity of the network to adopt new technologies, mainly in this context of the era of communication and the internet of energy, which requires a constant modernization of power systems.

## 3.2 DESIGN CHALLENGES AND PHILOSOPHICAL CONSIDERATIONS

This section aims to present the technical and system challenges in the SGs design. For this, the main technical barriers are raised to provide an overview that implementations and projects must have as a premise to find solutions. To complement this approach to the technical aspects of the grid, a holistic debate of the new context in which SGs are inserted, looking for highlight the impacts and how these changes relate to other systems and stakeholders.

### 3.2.1 CHALLENGES AND TECHNICAL BARRIERS

#### 3.2.1.1 Renewable Generation Sources

Several electric grid operators face this new challenge of integrating renewable sources into the grid, combined with the characteristic of distributed generation [9, 10]. The penetration of these sources has increase each year, in an impressive way, which makes the energy distribution and transmission systems need adaptations to host this new way of generating energy. The latest report [11] by the International Renewable Energy Agency (IRENA) with capacity statistics, confirms this growth. Figure 3.1 shows this statistics on the capacity of renewable energy capacity in the world.

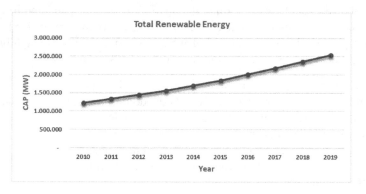

**Figure 3.1**   Total renewable energy world's capacity

A very significant feature of this type of generation is the intermittency of most sources. With an emphasis on solar PV and wind generation, which stand out in the distributed generation market and are widely used in various networks, including own use by the end consumer. Thus, with this immense number of generating units based on RES in the grid, there is a concern with the degradation of energy quality when adding this complexity. Figure 3.2 shows the percentage of growth of Wind and Solar PV generation in relation to the total capacity.

Another issue of concern regarding this large penetration of distributed generation is the injection of active energy at the level of medium voltage (MV) and low

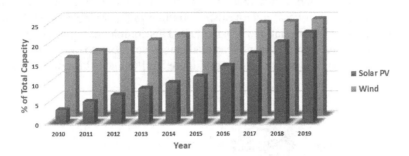

**Wind and Solar PV Generation**

**Figure 3.2** Percentage of Wind and Solar PV generation in relation to the total capacity

voltage (LV) level, this can result in overvoltage, especially when the R/X ratio is high as rural grids. Problems with unforeseen islanding and difficulty of protection systems to deal with this high number of generation units in the distribution network can also be a concern, because the design of previous grid was not foreseen for this size of insertion, which requires changes in strategies and infrastructure. In addition to these issues, the increasing use of renewable energies and associated electronic equipment can increase power quality problems, such as, for example, an increase in the harmonics emission and high-frequency emissions (2 – 150 kHz), called supra-harmonics.

A very important point is the balance between load and generation, therefore, there is a warning about this high penetration that designs and implementations must take into account in order not to overload the system. Therefore, there are studies that seek to investigate this capacity of electrical systems to host a certain amount of renewable generation without compromising the structure, reliability, and security of the network.

In transmission networks, this large number of new generation units may impact the creation of new power flows a concern due to the possibility of overload, already mentioned, and lack of operational reserves. As the function of the transmission system is to maintain excellent reliability, the aspects discussed in this section, such as generation uncertainty, an abrupt increase in generating sources, and changes in the operation profile can impair the performance of these networks. The Planning of the transmission system used to occur only for two extreme cases: maximum consumption and minimum consumption, in which the amount of production would change according to the amount of consumption [12]. According to this author, two more classes of operations appear that were missing previously, where extremes of consumption and generation are not found, further increasing the complexity of operations.

### 3.2.1.2  Management and Market Complexity

In this context of SGs, there is the integration of several elements including generation units distributed at different levels of the network, electric vehicles, energy storage, traditional network, microgrids, in addition to other elements related to the communication and information infrastructure. This complexity requires a more precise energy management that can present advanced solutions to deal with these different variables without losing efficiency. Figure 3.3 Ilustrates this complexity of the SGs.

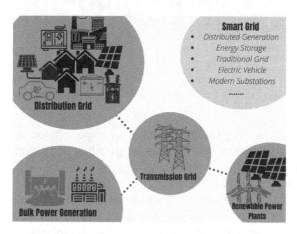

**Figure 3.3**  SGs elements and the management complexity

One of the biggest challenges is to coordinate a strategy that integrates all network infrastructures, be it traditional network or the modern elements network, and the modern elements that characterize SGs. The starting point is to maintain energy efficiency, avoiding losses and managing loads and generation correctly. These strategies must take into account the RES intermittency, network operating conditions, varied consumption, dynamic rates, among others. Energy efficiency is one of the main concern for sustainable development activities, because the increase in energy consumption usually implies on increase of $CO_2$ emissions and lasting impact on global warming [13].

It is important to emphasize that the communication systems are drivers for the integration of the different systems of SGs. Together with modern measuring equipment, such as smart devices and smart metering, they allow massive and bidirectional data transfer through wired and wireless communication protocols. Thus, the network as a whole becomes more capable to managing this range of information from many elements of generation, consumption, and operation. This challenge of dealing with all this complexity also involves data management, as many devices exchange information with an even greater participation by the end user in the operation.

The fundamental control problem is managing resources of bulk energy system (e.g. power generation resources) and dispatchable DERs in a way that will not compromise the operational requirements of grid, e.g. observing constraints on system frequency, voltages, and the operational limits of the grid components [14]. In this new grid format, this problem is increasingly aggravated by the characteristics mentioned above. Furthermore, the fact that distributed generation units are not, in most chaos, owned by the energy supply agency further complicates treatment energy traditional. The following is a list of the main challenges related to the new way of managing power systems.

1. Add a wide range of subsystems and technologies (bulk generation, distributed generation, transmission, distribution, substations, communication, energy storage, etc.);
2. Unpredictability of non-dispatchable energy sources, especially intermittent renewable sources;
3. Coordinate a massive bidirectional data flow;
4. System dynamics becoming faster and unexpected;
5. Better to integrate consumers, agencies, and prosumers;
6. Energy efficiency and reduction of $CO_2$ emissions;
7. Adapt and update the grids infrastructure;
8. Need to mature the bidirectional energy flow;
9. Reduce the peak load.

The electricity market in the context of SGs will largely depend on the demand response and energy efficiency programs [15]. In addition to several changes in the operation of electrical power systems, the way the market will be adapting also changes. In this context of SGs, other agents become active in the electricity market, for example the prosumers, who are the consumers that also generate their energy and surpluses. The market need to take this phenomenon into account and adapt the regulatory processes for effective participation and fit the use of new technologies. Demand Response mechanisms and incentives characterize basic objectives for utilities, business, and residential customers to reduce energy use during periods of peak demand, be efficient when power reliability is at risk, optimize the balance of power supply, and demand regardless of the system size [16].

This whole context addressed so far shows a change in the consumption profile and in the way the network operates to maintain efficiency and quality in energy supply. The market is challenged to face these transformations as an opportunity to employ new technology and approach. Perhaps the regulatory barriers are the most challenging in this process, as they need to break with rules that make the network work for a long time, but they are the most important for the development of a modern electricalgrid. Figure 3.4 highlights the challenges inherent of the complexity of managing SGs.

### 3.2.1.3 Power Quality Issues

The term power quality is related to voltage quality and supply continuity. The increasing complexity of the electric power network and its management, growing demand and expectations of service quality such as greater reliability, efficiency, and security of grid as well as environmental and energy sustainability

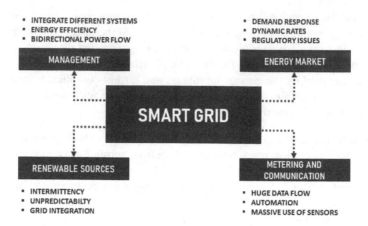

**Figure 3.4**  Challenges inherent of SG complexity

concerns triggered the next major step in the evolution of the electric power system toward a flexible network or SG [17].

As already mentioned in this chapter, the large insertion of RESs together with the widespread use of electronic equipment, such as PV inverters, cause impacts and power quality problems. In addition, demand side issues, poor resources coordination, and poor system integration can contribute to increase in network disturbance and problems at different voltage levels. Thus, it is possible to realize, to maintain high performance, SGs must present high performance on different sides of the operation. Table 3.1 details the quality problems, definitions, and relationship with the new network model.

The table illustrates some quality problems, but the challenges are not limited to them. Problems in network operation, interruption, are also happening and can generally be associated with contextual changes for the SG.

An important concept associated mainly with distributed generation is Hosting Capacity. This investigation aims to verify how much generation can be associate with the system without exceeding the acceptable deterioration limits. For the SGs design, this step is of paramount importance. Figure 3.5 illustrates the concept. When it comes to recent problems, such as supraharmonics, greater care must be taken due to the lack of information and rates to qualify. Thus it is necessary to seek new technological knowledge of power quality.

The use of advanced signal processing techniques, artificial intelligence, big data, and other types of computational analysis, is a good opportunity to overcome the obstacles presented, together with driving technologies that become powerful in these aspects.

Finally, interruption events caused by network incidents or even natural catastrophes, are undesirable and become a challenge for the current electrical system and society as a whole. This type of event should be viewed more carefully.

**Table 3.1**
**Power Quality Issues Associated with SGs**

| Quality Issue | Short Definition | Example |
|---|---|---|
| Overvoltage | An increased in the effective voltage greater than 1.1 pu for a period greater than 1 min (1.1 pu to 1.2 pu). | Solar panels connected to low-voltage networks result in over-voltages. |
| Voltage Fluctuations | A series of voltage changes or a cyclical variation of the voltage envelope. | Caused by wind turbines and photovoltaic generators due to the fluctuations in wind speed and solar radiation. |
| Harmonics | Harmonics are sinusoidal voltages or currents with frequencies that are integer multiples of the frequency at which the supply system was designed to operate (called the fundamental 50 Hz or 60 Hz). | Even more harmonic emission from electronic inverters/converters, EV Chargers, etc. |
| Interharmonics | A frequency component of a periodic quantity that is not an integer multiple of the frequency at which the supply system is operating (e.g. 50 Hz, 60 Hz). | Wind generators converters also emit interhamonics. |
| Supraharmonics | High-frequency components in 2-150 kHz range. | High switching PV inverters emit this type of disturbance. |
| Flicker | The subjective impression of the floating luminance caused by voltage fluctuations. | Repeated start-up of heat pumps may result in visible light flicker. |

*Source:* Definitions from "IEEE Recommended Practice for Monitoring Electric Power Quality" in IEEE Std 1159-2019, vol., no. (2019): 1-98.

*Note:* Listed are the most common phenomena that are associated with Smart Grids.

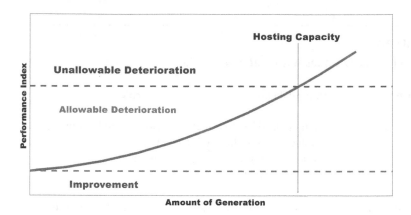

**Figure 3.5**  Hosting capacity concept (Adapted from [17])

In the context of a flexible system, the network must be prepared to present quick and safe solutions in the face of gravity. This agreement is very important and is connected to the self-healing characteristic that smart systems must-have.

### 3.2.1.4  Cybersecurity

The SG is a network of computers and power infrastructures that monitor and manage energy use [19]. Each application, equipment in operation, measuring devices spread by private and public entities, is integrated into the electrical information sharing system. This arrangement raises concerns about information security, privacy, and malicious activity on the system. In addition, there is still the possibility of fraud, leakage of private consumer information, changes in consumption data.

Cyber security objectives can be classified into the following three categories [20, 21, 22]:

1. **Integrity:** Protection against the unauthorized modification or destruction of information. Unauthorized access to information opens the door to misuse of information, leading to mismanagement or misuse of power.
2. **Confidentiality:** Protect privacy and proprietary information by authorized restrictions on access and disclosure.
3. **Availability:** Ensure timely and reliable access to information and services. Availability can be compromised by disrupting access to information that impairs the power supply.

The proposed solutions include encryption, aggregation of data at high levels to mask individual user, limiting the amount of data to only the information needed for billing purposes and real-time monitoring of these networks [23, 24]. Still, malicious intruders can be sophisticated enough to circumvent security. Keeping the power grid and the information network secure should be a priority, as the way the system will integrate in the future makes it more vulnerable and susceptible to attacks. The use of machine learning can be an ally in combating digital terrorism in SGs and in the secure communication of devices.

## 3.2.2  HOLISTIC NORMATIVE ENGINEERING DESIGN FOR SMART GRIDS

The great complexity of SGs requires several solutions to the challenges of system integration, in addition to dimensioning the impacts for different stakeholders. The propose of this approach is to draw attention to Normative Design Frameworks, which attempts to cover the complexity of the grid on the future taking into account the most important components and relationships in an integrated holistic way.

The Philosophy of technology is like a mosaic of ideas and suggestions, but this should not minimize its relevance to engineering and technological developments [25]. The ongoing transformations of the network of the future, which are becoming increasingly intelligent, raise a number of technical, social, and economic issues that cannot be addressed separately. It is important that technological project developments are aligned with all the faces and impacts that the final deployment is influencing.

The changing scenario has promoted the development of new concepts in which SGs have become the new design approach for the development of the future electric networks, allowing integrated and enhanced performance and diagnosis [26].

In [27], the levels of complexity of SGs can be distinguished between three strongly intertwined categories: dimensional complexity, technological complexity, and stakeholders complexity. Figure 3.6 illustrates this division.

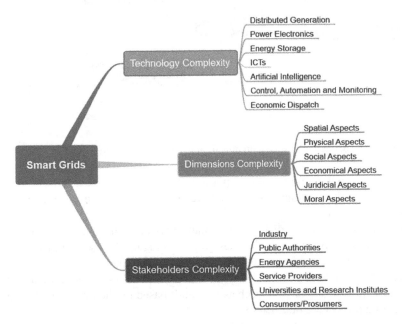

**Figure 3.6**  Framework of complexity in SGs

Dimensional complexity has to do with the various dimensions in which SGs are involved. Complexity Technology is linked to the technologies implemented to promote this network format. The Stakeholder complexity is based on the interests of those involved and the promoters.

The view of Engineering Systems is essential to play this context in the design process. Engineering Systems blend engineering with perspectives from management, economics, and social science to address the design and development of complex, large-scale, sociotechnical systems. The four underlying subfields of this approach are:

1. Systems Engineering (including systems architecting and product development);
2. Operations Research and Systems Analysis (including system dynamics);
3. Engineering Management;
4. Technology and Policy.

The Systems Engineering subfield is a powerful tool for analysis that is focused on the total operation of the system. According to [28], it looks at systems from the outside, that is, at its interactions with other systems and its environment, as well as from the inside. Figure 3.7 shows the relationship between concepts.

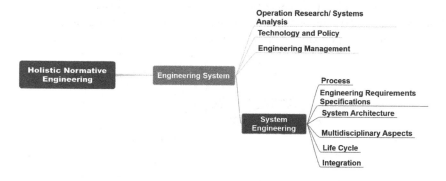

**Figure 3.7**   Expanded relationship in holistic normative engineering

In addition to these points, it is important to establish design criteria, which must meet the normative principles. Figure 3.8 illustrates the aspects that should be related to the design criteria.

In [29] and [30] a key distinction is made between "dimensions" and "things". This theory is important so that the design process is not focused on just the formative aspect. That view shows that the systems are linked to different modal aspects or different dimensions. So, it can be applied as a list of tasks, where critical thinking must raise questions related to each aspect, and, based on their answers, the specifications of a project are outlined [31].

According to [32] there are four properties in this type of analysis. First, each dimension has its nature and cannot be attributed to another. Second, any

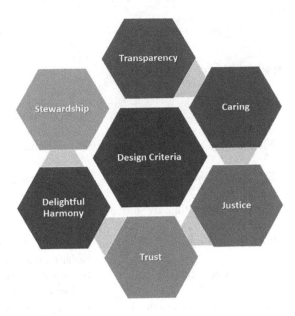

**Figure 3.8**  Design criteria aspects

technology or technology system can be distinguished by the 15 modal aspects. Third, the modal aspects have a propagation characteristic, in other words, one aspect influences the other. Finally, a distinction can be made between the types of aspects, between natural (arithmetic to biological) and behavioral (psych to pistic).

SGs can be analyzed from the perspective of these modal aspects. From the design point of view, this is an important step in understanding and listing the dimensions involved in the context as a whole. Table 3.2 shows how each modal aspect is also related to SGs.

Through these tools, it is possible to build a holistic normative design for the future of power systems. In this way, the technologies and the electrical network itself will be able to deal with the technical and multidisciplinary challenges in a more efficient and comprehensive method, in addition to having the possibility of being centered on human development. Figure 3.9 summarizes the framework of the approach.

## 3.3  SMART GRID ARCHITECTURES AND TECHNOLOGIES

This section aims to highlight the main architectures that characterize SGs, together with the main technologies that drive power systems in terms of advances in quality, reliability, and smartness.

**Table 3.2**
**Modal Aspects Applied to SGs**

| Modal Aspect | Core Aspect | Application for SG |
|---|---|---|
| Arithmetical | Numbers | Measureable quantities: voltage, current and power. |
| Spatial | Extent, unbroken extent | Spatial arrangement of transmission and distribution lines. |
| Kinematic | Movement, continuous movement | Rotating machines, energy flow. |
| Physical | Energy, interaction | Properties of conducting and isolating materials. |
| Biotic | Life, organic, vegetative, vital | Influence electromagnetic fields and waves on life. |
| Psychic | Feeling, sensitive, sensorial | Feelings of safety and control of humans in a smart environment. |
| Analytical | Logic, rational, analytical distinction, conscious distinction | Distinction between different types of grids: micro, smart, super, etc. |
| Formative | Controlled forming, power of freedom, power, domination | Control of power generation, distribution and consumption, smart meters. |
| Lingual | Denotation, meaning, symbolic meaning | Term "smart" chosen to promote technology? Should it be smarter? |
| Social | Intercourse, coherence, communion, interconnectedness | Influence of micro grids and SGs on the behavior of and interaction between users. |
| Economic | Control of rare goods, stewardship, fertility, productivity | Price differentiation depending on momentary supply and demand. |
| Aesthetic | Harmony, beauty, allusion, full diversity of shades | Beauty of urban grid structures. |
| Juridical | Retribution, justice, law | Who is liable for a failing SG? When are SGs appropriateness and when not? |
| Moral | Love, care, fidelity, willingness to serve | Contribution of SGs to a sustainable future. Safety of energy generation and transport. |
| Pistic | Transcendental certainty, reliability, faith, credibility | Vision/commitment/belief on a smarter grid. |

*Source:* Adapted from Paulo Ribeiro, Henk Polinder, Maarten Verkerk, "Planning and Designing Smart Grids: Philosophical Considerations, *Technology and Society Magazine, IEEE* 31 (2012): 34-43.

*Note:* These are the dimensions of SGs.

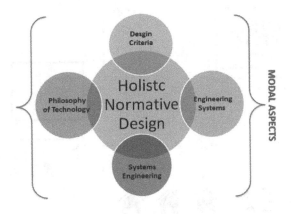

**Figure 3.9**  Framework for holistic normative design

## 3.3.1   THE COMMUNICATION STRUCTURE AND TECHNOLOGIES

The ICTs have already been presented as one of the major drivers for the development of SGs. The trend of electric power systems is to present communication at all levels of the network, in addition to the present one with high performance in the generation, transmission, and mainly in distribution. These technologies assist in the coordination of network systems, providing an exchange of bidirectional information, facilitating management, and facilitating data exchange between agents and various equipment [33]. In the context of distributed generation and the greater need to manage the generation resources that are also present in distribution, the key role of this type of technology is more evident.

In terms of communication, the electrical networks are based on the Wide Area Network (WAN), consisting of central control systems that facilitate the operation and traffic of information. With the advances in ICTs and the emergence of other concepts in the energy supply chain, other forms of communication have been adapted or incorporated.

In this way, the communication systems of the electric grid can be used in two more ways, the Neighborhood Area Networks (NAN) and the Home Area Networks (HAN). These network expansions are mainly linked to the new structure with distributed resources and consumer participation, in addition to a bidirectional flow of information and energy. Figure 3.10 shows the communication infrastructure of the SGs.

The main difference between these networks is in the coverage and transmission rate. HAN operates for residential or small applications with low data transmission (bps). Already, NAN operated in a larger amplitude covering distribution areas served by a substation for example, with a higher transmission rate (Kbps). Finally, WAN covers kilometers in length and has a higher transmission rate (Gbps). Moreover, the communication of all SG's components including operator control center, main and renewable energy generation, transmission and distribution, is based on WAN [34]. Table 3.3 shows the main technologies used in each sub-network.

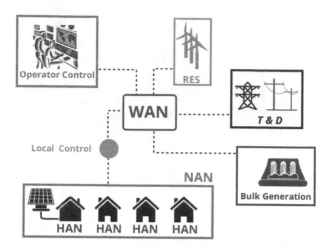

**Figure 3.10** Communication infrastructure

## 3.3.2 SMART METERING, MEASUREMENTS, CONTROL, AND AUTOMATION

The evolution of the power grid is moving toward an advanced technological level when it comes to solutions for measurement and automation. Smart applications require an improvement in these aspects, for example, perhaps one of the main ones demand-side management requires a dynamic consumer measurement that meets the context of distributed generation. For this reason, it is extremely important to have an architecture with a high level of automation and with bridge measurement technologies, which is also linked to communication technologies.

Traditional electric meters measure the total consumption of the end-user or are used to measure the amount of energy transported. Traditionally, accumulation meters are used, but in recent times, modernization has been adopted in this regard. Some modernizations were performed to allow users who had more control over their consumption and also to deal with different types of electric tariffs. These advances have allowed us to reach those known today as Smart Meters, which play a fundamental role in the formation of an SG.

Smart meters are even more sophisticated as they have two-way communications and provide a real-time display of energy use and pricing information, dynamic tariffs and facilitate the automatic control of electrical appliances [35]. Through an Advanced Metering Infrastructure (AMI), it is possible to guarantee the widespread use of this technology. This infrastructure is based on the communication structure previously presented. The Smart Meter is at the HAN level installed at the consumer and follows the communication chain. This data from several meters are grouped by the Local Data Aggregator, which it collects through the NAN. In the WAN, the control operator has access to the data emitted by the advanced measurement applications. Figure 3.11 shows the AMI scheme.

**Table 3.3**

**Technologies Applied to Each Sub-Network**

| Sub-Network | Communication Technologies |
| --- | --- |
| HAN | Ethernet, Wireless Ethernet, Power Line Carrier (PLC), Broadband over Power Line (BPL), ZigBee |
| NAN | PLC, BPL, Metro Ethernet, Digital Subscriber Line (DSL), EDGE, High Speed Packet Access (HSPA), Universal Mobile Telecommunications System (UMTS), Long-Term Evolution (LTE), WiMax, Frame Relay |
| WAN | Multi Protocol Label Switching (MPLS), WiMax, LTE, Frame Relay |

*Source:* Data from R. Bayindir, Ilhami Colak, Gianluca Fulli, Kenan Demirtas, "Smart grid technologies and applications", *Renewable and Sustainable Energy Reviews,* 66 (2016): 499-516.

In addition to adding a lot to demand response programs and bringing several benefits to users, these meters also allow increased resolution of data on various measurement parameters across the grid and these data can be used by utilities for the following applications [36].

1. Faster outage response plan by providing data to the field operations timely;
2. Better inform the consumer about the power grid status and progress in cases of an outage;
3. Improve the resilience of the grid by reducing interruptions and duration of the inevitable ones, improving the precise measures, and managing the responses to these events.

In sequence, the generation and transmission level are based on the Supervisory Control and Data Acquisition (SCADA) for the monitoring and control layer. Common components of a SCADA system are a human-machine interface, a supervisory system, a communication network, remote terminal units (RTUs), I/O devices, and control devices [37]. In this arrangement, I/O devices are sensors and actuators, whereas the control devices are the Programmable Logic Controllers (PLC) and Intelligent Electronic Devices (IED). The SCADA systems are also used to serve the distribution network. However, in this context with high integration of renewable sources, real-time operations, and quick response, these systems are not enough [38].

The smart metering and measurement devices that are advanced with recent developments in SG applications are Phasor Measurement Units (PMUs), IEDs, and

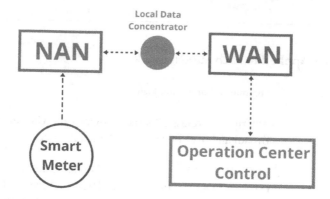

**Figure 3.11** The AMI scheme

smart meters [34]. The IEDs are used to detect faults, protection relaying, event recording, measurement, control, and automation aims in power network [39, 40]. PMUs play an important role on Wide Area applications and can be a key technology to increase accuracy and obtain better response times in energy management systems (EMS). This technology has a data entry and acquisition interface and a microcontroller, and the global positioning system (GPS) receiver and internal phase-locked loop (PLL) ensure synchronization of the entire system with universal time-coordinated (UTC) timestamp [34].

The EMS are automation systems that collect energy measurement data from the field and making it available to users through graphics, online monitoring tools, and energy quality analyzers, thus enabling the management of energy resources [41]. This system uses SCADA application to perform data analysis and monitoring. Among the functions are topological analysis, load forecasting, power flow analysis, and state estimation. With the evolution and consolidation of SGs, SCADA will need to incorporate new smart metering elements, such as phasor measurement units (PMUs) and intelligent relays, and consider new renewable sources of power generation and energy storage in electric vehicles (EV) [42]. Figure 3.12 shows a typical SCADA architecture.

A wide area monitoring system (WAMS) represents an important and promising technological advance in monitoring and control of actual and future power transmission systems [40]. This system aims to complement the SCADA application in transmission networks. This system is based on PMUs, and is then also made up of phasor data concentrators (PDCs), a communication network, and a power system communication center. WAMS allows monitoring, state estimation, and analysis of power system events. Figure 3.13 shows the architecture of WAMS.

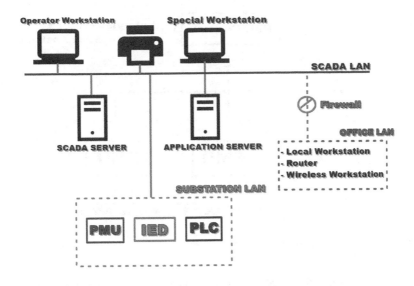

**Figure 3.12**   Typical SCADA architecture based on [40]

### 3.3.3   MICROGRIDS AND KEY TECHNOLOGIES

The microgrids can be characterized as active local distribution systems, with elements of distributed generation such as solar, wind, microturbines, diesel generator, fuel cell, among other forms of generation and energy storage technologies. Those systems can operate connected to the main power grid or in island mode, disconnected from the utility grid. In the connected mode, microgrids provide better integration of renewable sources, in addition to promoting control and management strategies that improve energy quality and consumer spending. In times of interruption of supply by the concessionaire, for example, the micro-network can start the operation in island mode and isolate itself from the main network, becoming a network with reliability and resiliency. The isolated operation of the microgrid also seeks to serve remote locations, those that are unable to connect to a bulk electrical network. In the latter case, they require specific approaches to guarantee safe and quality operation, in addition to these networks being extremely important for remote communities.

Associated with renewable sources and smart microgrid technologies are the network's electronic components, as power electronic converters that play a fundamental role in this context of distributed generation and for improvement of the power systems. Associated with renewable sources and smart microgrid technologies are the network's electronic components, as power electronic converters that play a fundamental role in this context of distributed generation and for improvement of the power systems. Voltage Source Converters (VSC) and

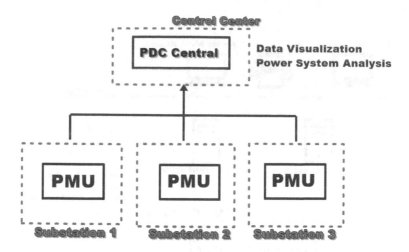

**Figure 3.13**  WAMS architecture

Current Source Converters (CSC) are the main electronic interfaces for a renewable energy source, especially when we are talking about PV energy, it is also associated with the use of energy storage systems. Other components that play important roles are the Flexible AC Transmission Systems (FACTS), these are used to increase the power transfer capability of existing AC lines, to control steady-state and dynamic power flow through an AC circuit, to control reactive power and voltage and to enhance voltage and angle stability [43].

Microgrid architectures can be classified into three groups: AC microgrid, DC microgrid, hybrid AC/DC microgrid. The AC microgrid has a common AC bus which is generally connected to mixed loads (DC and AC loads), distributed generations, energy storage devices [44]. This has great compatibility with traditional AC networks, in addition to having greater reliability and capacity. On the other hand, when it comes to DC loads and sources, the integration is done by DC/AC inverters, which reduces efficiency [45, 46]. Figure 3.14 shows an example of AC architecture scheme.

In DC microgrid, common DC bus is used to connect to the grid through an AC/DC converter [44]. The DC microgrid has some advantages, one of which is to avoid losses due to energy conversion, it also has better integration with renewable energy sources. These also have greater stability. Finally, we have the hybrid AC/DC microgrid, which is the combination of the two architectures, presenting a sum of benefits and advantages in use. Figure 3.15 shows an example of DC architecture scheme.

Energy storage plays a key role in microgrids. This technology can be used in different ways, one of which is to use a large capacity to support the microgrid in reducing fluctuations and improve stability. Energy Storage also performs some energy management functions, such as load leveling, peak shaving, and RES energy shifting.

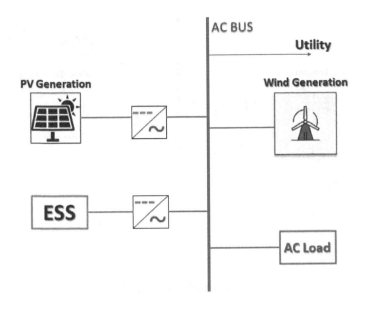

**Figure 3.14**  Example of AC architecture scheme

It is possible to use energy storage also with the cargo or with the generation. Along with the load, it provides support when variations are obtained that require a quick response in which the generation follows the change in the load, thus avoiding sudden operations. Another way is to use storage with renewable generation to smooth the output of these intermittent sources. Below is a list of the main technologies in the sector. Figure 3.16 shows the applications and used technologies. The list below includes some of the most used technologies.

1. **Battery:** Battery is an electrochemical storage solution and the most applied in the market. Examples of types are the lithium-ion (Li-ion), lead-acid, nickelcadmium (NiCad), sodium/sulfur (Na/S), vanadium-redox;
2. **Flywheels:** Mechanical devices that take advantage of rotational energy to provide instant electricity;
3. **Compressed air energy storage:** Using compressed air to create a reserve of potential energy;
4. **Thermal storage:** Capturing heat and cold to create energy on demand;
5. **Pumped hydro storage:** Big reservoirs of water to store energy;
6. **Superconducting magnetic energy storage (SMES):** In SMES, the energy is stored in the magnetic field of a superconducting coil formed by direct current flowing through the coil.

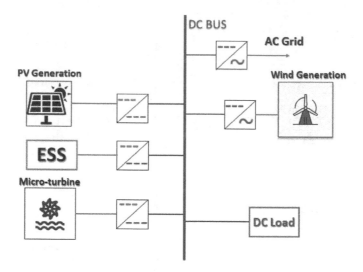

**Figure 3.15**  Example of DC architecture scheme

## 3.4  INTEROPERABILITY AND SCALABILITY

### 3.4.1  MOVING FOR INTEROPERABILITY IN SMART GRIDS

The evolution of the energy sector has the main characteristic of being digital. Changes in the way of generating, distributing, and consuming energy go through a modernization process that is made possible by the wide use of ICTs, applications are no longer limited to electrical networks, they are now connected to the consumer and to new models of the grid.

A crucial point to make it increasingly possible to search for a digital electric sector is to move toward interoperability. It is the characteristic that allows two types of equipment or systems to operate together without any conflict of communication or compatibility. This is a very big challenge for power systems because each part of the energy supply chain has its standards, protocols, and culture to deal with its equipment operations. For this to occur, there must be a policy effort and convergence between institutional standards for the electricity sector.

The benefits related to interoperability vary among stakeholders of SG design, but everyone benefits from better clarity between operations, greater affordability, among other advantages. In this process, there are some difficulties to be overcome, following a list of key issues for this process [47].

1. The large number of stakeholders, different considerations, number and complexity of standards available (and missing) requires a more formal nationally driven governance structure;
2. Since SG efforts are underway, and in some cases complete, standards adoption must consider work already completed and underway;
3. Interoperability discussions and definitions should be expanded to focus on standards across systems (inter-system) rather than just within systems (intra-system);

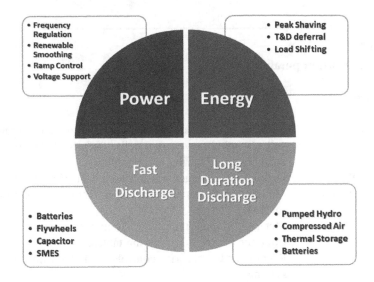

**Figure 3.16**  Technologies and services of energy storage for microgrids

There is a set of standards for the SG that has been proposed, the Institute of Electrical and Electronics Engineers (IEEE) develops standard series that are related to SG interoperability and also to the interconnection of distributed resources such as IEEE 2030 and IEEE 1574, respectively [48, 49]. Other IEEE standards that are related to communication elements also contribute to the progress of interoperability, these are: IEEE 802.3 (Ethernet); IEEE 802.11 (WiFi); IEEE 802.15.1 (Bluetooth); IEEE 802.15.4 (ZigBee); IEEE 802.16 (WiMax).

The International Electrotechnical Commission (IEC) also proposes a standard that has been applied to the electricity sector The American National Standard Institute (ANSI), National Institute of Standards and Technology (NIST), among others, also present a convergence of standards to promote interoperability. Table 3.4 shows a list of standards and the key applications.

For the success of SGs to be achieved, it is necessary to create an impulse for interoperability, the development of standards and technologies that reconcile the interests of all agents is essential to achieve the objective.

## 3.4.2   SCALABILITY ASPECTS FOR THE MODERN GRID MODEL

Scalability can be defined as the ability of a system to maintain its performance (i.e. relative performance) and function, and retain all its desired properties when its scale is increased without having a corresponding increase in the system's complexity [50]. This evaluation is extremely important to provide support to reduce barriers to development, in addition to accessing projects that have an estimate of the

**Table 3.4**

**Standard for Interoperability**

| Standard | Application |
|---|---|
| IEC 61850 | Substation automation, distributed generation (photovoltaics, wind power, fuel cells, etc.), SCADA communications, and distribution automation. |
| IEC 61968 | Distribution management and AMI back office interfaces. |
| IEC TC 13 and 57 | Metering and communications for metering, specifi cally for AMI. |
| ANSI C12.19 | Metering "tables" internal to the meter. |
| ANSI C12.22 | Communications for metering tables. |
| NIST SP-800.53 | Recommended security controls for federal information systems. |
| NIST SP-800.82 | Guide to Industrial Control Systems (ICS) security. |
| NERC CIP 002-009 | Bulk power standards with regard to Critical Cyber Asset Identification, Security Management Controls, Personnel and Training, Electronic Security Perimeter(s), Physical Security of Critical Cyber Assets, Systems Security Management, Incident Reporting and Response Planning, and Recovery Plans for Critical Cyber Assets. |

*Source:* Data from James Momoh, "Interoperability, Standards, and Cyber Security", in Smart Grid: Fundamentals of Design and Analysis, IEEE (2012), pages 160-175.

technical and economic viability of the application. Some factors characterize the analysis, they follow below [51].

1. Technical factors indicate the feasibility of expanding a project. It can guarantee that a solution is reproducible;
2. Economic factors reflect whether an expansion is worth the investment. This step seeks to validate the business model for different scales or alternative scenarios to the original case, and cannot be neglected;
3. Factors related to regulatory regulation will define the potential for acceptance and whether stakeholders are ready to adopt an expanded or alternative version of the original project.

Technical factors are linked to the modularity of the application, as it evolves and relates to other systems. For example, automation projects in microgrids should easily be able to adopt different control and protection schemes, as needed, as well as being able to add or remove components without impairing system performance. Technical factors are linked to the modularity of the application, as it evolves and relates to other systems. For example, automation projects in microgrids should easily be able to adopt different control and protection schemes, as needed, as well as being able to add or remove components without impairing system performance. The application must be friendly to the professionals of the environment, that is, it must be comprehensible to different engineers and technicians. Finally, to be compatible with the maximum possible systems, this will be decisive in the project's expansion process.

The economical factor is to understand if the project or application is economically viable. In an expansion context, the economic return has to accompany the change in level, that is, remain viable. The economical factor is to understand if the project or application is economically viable. In an expansion context, the economic return has to accompany the change in level, that is, remain viable. In this case, one should use metrics that guarantee the analysis of economic projections. It is something analogous to the idea of Hosting Capacity when I can expand without harming the profits of the project.

The regulatory process defines the interaction between agents and stakeholders. In the energy sector, it will coordinate the relations between consumer, agency, institutions, operators of the electrical systems, among others. The transmission, distribution, and power generation systems also go through the regulatory process. The rules bring limits to scalability, for example in Brazil, where the treatment regime for distributed generation is still under debate because the current regime does not allow the sector to expand and the regulator seeks better solutions to try to promote the best for all concerned. Thus, the SG project must consider the regulation of where it is being implemented. Although the solution proposed by a project might have overcome the regulatory and legal barriers (e.g. by adapting the regulatory framework), it is very important that other stakeholders accept the proposed solution [51].

In addition to taking these factors into account, one must be concerned with using tools, identifiers and parameters for the scalability analysis. Some analysis are proposed in [52, 53]. Figure 3.17 illustrates the factor for achieve the scalability.

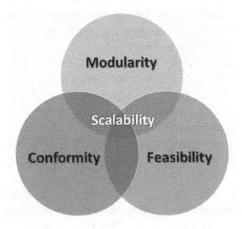

**Figure 3.17** Factors to achieve scalability

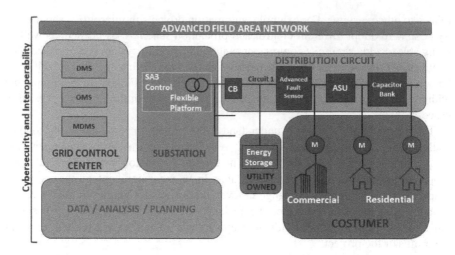

**Figure 3.18**   IGP scope from Southern California Edison (adapted from [55])

## 3.5   APPLICATIONS

This last section covers some applications of SGs, presenting some examples to create a context of understanding. It is important to understand the approaches proposed in this chapter from a practical project perspective, whether in the form of research or implementation.

### 3.5.1   DISTRIBUTED ENERGY RESOURCES MANAGEMENT

The Distributed Energy Resources (DERs) management is a leading issue to guarantee reliability and quality in distribution systems. Due to the significant insertion of distributed renewable sources, coordination became more complex due to the intermittent nature of these sources [54]. The widespread use of metering associated with communication technologies has become a great ally in this process, in addition to control strategies driven by key technologies.

The energy concessionaire, Southern California Edison, has a project called Integrated Grid Project (IGP) [55], in California, United States of America (USA) whose main objective is to ensure effective management of the DERs. The company is one of the pioneers and encourages the use of clean energy to reduce $CO_2$ emissions and presents several projects aimed at the broad use of advanced technologies in electric grids. It is also a goal of the project, to investigate potential architectures on the network, and to promote an optimal operation with a high level of penetration of distributed resources.

The IGP is focused on optimizing all DERs assets of third parties or these owned by the utility. Thus, the project is organized into three sub-applications. The first one is the control of the DER systems, basically voltage control and power flow optimization. Another part of the project is focused on integrating the system via ICTs. Lastly, the project aims to aggregate participating users to hire their private distributed generation. Figure 3.18 shows the scope of project.

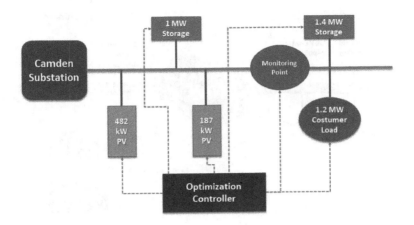

**Figure 3.19**  IGP control scheme

Thus, the project hosted two areas managed by two substations (Camden and Johanna Jr.) to carry out the project. This choice was because it already has large PV installations, in addition to engulfing the profile of residential and commercial consumers. Some observations on the benefits can be seen below. Figures 3.19 and 3.20 show the control scheme and high level logical architecture, respectively.

1. The control of the DERs increases the hosting capacity due to the optimization of the power flow and voltage control;
2. The operational service bus provides easy integration of different applications through standardization;
3. The field area network enables low latency in the communication of advanced automation applications;
4. Implementation of the IEEE 2030.5 standard enables a method of communication between smart inverters and aggregators.

This is an example of an SG application carried out by utility that brings several benefits to the community and the local energy sector as a whole. It presents advanced concepts of control and integration of renewable sources, as well as integration through communication and standardization technologies.

### 3.5.2  ENERGY STORAGE

Energy storage not only plays an important role in microgrids but also in other diverse applications in SGs. To propose a more intelligent, sustainable network, with a high level of penetration of renewable sources combined with quality and reliability, it is not possible to move forward without using storage technologies.

The Public Service Company of New Mexico and its partners played a project [56, 57] that was an application of energy storage to support the operation of PV

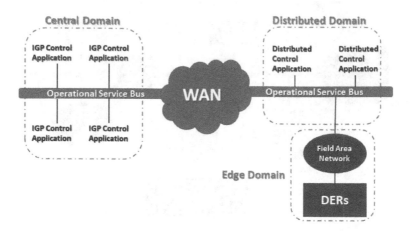

**Figure 3.20** IGP high-level logical architecture

generators. A 500 kW/1 MWh advanced lead-acid battery was used in conjunction with separately installed 500 kW solar photovoltaic (PV) plant to create a dispatchable distributed generation resource. The site is situated in southeast Albuquerque, EUA.

This arrangement provided voltage smoothing and peak shifting through the use of advanced control algorithms. Data analysis also contributed to other applications and the optimization of the battery systems. The objective was to highlight the role of using energy storage with intermittent sources, showing that it is possible to reduce fluctuations, among other problems. There was also the intention to make a reduction of at least 15% of the peak load viable. Figure 3.21 shows the One Line electrical scheme of the complete system.

That is a typical application that highlights the role of battery systems and storage technologies in general. Among the benefits achieved by the application, it is possible to highlight the reduction in electricity costs, increased network efficiency, strengthening energy security, and reduction of $CO_2$ emissions.

### 3.5.3   METERING AND AUTOMATION

The application of new measurement technologies supports advances in other issues in the network. A large part of the structuring of Smart Grid will be supported by this equipment. The PMU is one of those technologies that gains more and more space in the sector and brings great benefits to the power systems.

Duke Energy's conducted a PMU deployment project [58] in North Carolina and South Carolina, USA. This implementation was only about a plan to modernize communication systems by installing PMUs and a phasor data concentrator. Several 230 kV and 500 kV substations have been upgraded. This technology allows for better measurement, which leads to improved management and control, making the power

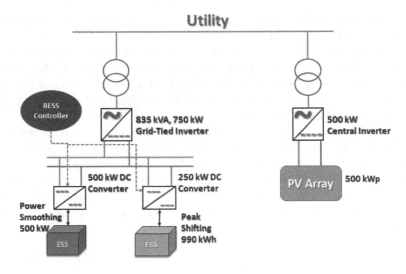

**Figure 3.21** Public Service Company of New Mexico's Project Architecture

grid more reliable. Below are the possible applications in advanced transmission for this technology.

1. Angle and frequency monitoring to provide better information on the network and power flow;
2. Modern analyzer for large-scale investigation of events, that is, trying to measure grid bottlenecks;
3. Aggregation in state estimation, improving models, planning, and operation;
4. Increases the accuracy of steady-state model through the benchmarking models.

The benefits of this project include improving the reliability of transmission systems, implementing the ability to better understand events, promoting improvements in communication systems for transmission. Finally, the project also prepares the transmission systems to receive DERs.

## 3.6 SUMMARY

The introduction (Section 3.1) provides an overview and contextualization of the subject, presenting the main starting points so that there is an understanding of the content detailed in the chapter.

The following section (Section 3.2), which addresses the main challenges for smart grid projects, details these challenges at different levels of the power systems of the future. Challenges inherent to the complexity of the electrical network in terms of the market, renewable integration sources, quality problems, and cybersecurity are addressed. Also, it brings debate and essential philosophical concepts to evaluate and promote Smart Grids projects.

Also presented are the main architectures and the key technologies (Section 3.3) to achieve the Smart Grid model. This section discusses how these technologies are associated with new structure frameworks, which in turn go back to the new paradigms of power systems. This section is divided into three fields. These are communication structure, automation and metering, and microgrid and distributed generation.

In the penultimate section (Section 3.4), the main aspects of interoperability and scalability for projects and smart grids, are addressed. At this point, the text points out the main standards that address these topics and evaluation aspects.

Finally, some Smart Grid applications (Section 3.5) are presented, collected from real projects using technologies and concepts covered in the chapter. These projects are described and illustrated to contextualize the theory with applications already implemented.

## REFERENCES

1. Kumar, N., Aujla, G.S., Das, A.K., and Conti, M. 2019. ECCAuth: A secure authentication protocol for demand response management in a smart grid system. *IEEE Transactions on Industrial Informatics,* 15(12), 6572-6582.
2. Kaur, D., Aujla, G.S., Kumar, N., Zomaya, A.Y., Perera, C., and Ranjan, R. 2018. Tensor-based big data management scheme for dimensionality reduction problem in smart grid systems: SDN perspective. *IEEE Transactions on Knowledge and Data Engineering,* 30(10), 1985-1998.
3. Lopes, J.A.P., Madureira, A.G., and Moreira, C. 2019. A View of Microgrids. In *Advances in Energy Systems: The Large-scale Renewable Energy Integration Challenge,* ed. P.D Lund, J. Byrne, R. Haas, and D. Flynn, 149-166. Hoboken: John Wiley & Sons.
4. Jindal, A., Kumar, N., and Singh, M. 2020. Internet of energy-based demand response management scheme for smart homes and PHEVs using SVM. *Future Generation Computer Systems,* 108, 1058-1068.
5. Jindal, A. 2018. Data Analytics of Smart Grid Environment for Efficient Management of Demand Response. PhD Thesis. Thapar Institute of Engineering and Technology, Patiala.
6. IEA. 2020. Tracking Energy Integration. Technical Report, IEA, Paris. https://www.iea.org/reports/tracking-energy-integration (accessed April 10, 2020).
7. Chaudhary, R., Aujla, G.S., Kumar, N., Das, A.K., Saxena, N., and Rodrigues, J.J. 2018, May. LaCSys: Lattice-based cryptosystem for secure communication in smart grid environment. *In 2018 IEEE International Conference on Communications (ICC)* (pp. 1-6). IEEE.
8. Aujla, G.S., Garg, S., Batra, S., Kumar, N., You, I., and Sharma, V. 2019. DROpS: A demand response optimization scheme in SDN-enabled smart energy ecosystem. *Information Sciences,* 476, 453-473.
9. Aujla, G.S., Singh, M., Kumar, N., and Zomaya, A. 2017. Stackelberg game for energy-aware resource allocation to sustain data centers using RES. *IEEE Transactions on Cloud Computing.* 7(4), 1109-1123.
10. Kumar, N., Aujla, G.S., Garg, S., Kaur, K., Ranjan, R., and Garg, S.K. 2018. Renewable energy-based multi-indexed job classification and container management scheme for

sustainability of cloud data centers. *IEEE Transactions on Industrial Informatics,* 15(5), 2947-2957.

11. IRENA. 2020. Renewable capacity statistics 2020 International Renewable Energy Agency. Technical Report, IRENA, Abu Dhabi. `https://www.irena.org/publications/2020/Mar/Renewable-Capacity-Statistics-2020` (accessed April 10, 2020).

12. Bollen, Math. 2011. *The Smart Grid: Adapting the Power System to New Challenges.* San Rafael: Morgan & Claypool.

13. Miceli, Rosario. 2013. Energy management and smart grids. *Energies,* 6, 2262-2290.

14. DOE. 2018. Smart Grid System Report 2018. Technical Report, United States Department of Energy. `https://www.energy.gov/oe/downloads/2018-smart-grid-system-report` (accessed April 20, 2020).

15. Lin, Jeremy and Magnago, Fernando. 2017. Electricity Market under a Future Grid. In *Electricity Markets: Theories and Applications: Theories and Applications*, ed. J. Lin, and F.H. Magnago, 293-314. Wiley-IEEE Press.

16. Torres, Juan J. 2011. Smart Grid: Challenges and Opportunities. Technical Report, United States Department of Energy. `https://www.osti.gov/servlets/purl/1120614` (accessed April 17, 2020).

17. Zavoda, F., Rönnberg, S., Bollen, M., et al. 2018. *Power Quality and EMC Issues with Future Electricity Networks.* CIRED.

18. "IEEE Recommended Practice for Monitoring Electric Power Quality," in IEEE Std 1159-2019 (Revision of IEEE Std 1159-2019), 1-98, doi:10.1109/IEEESTD.2019.8796486.

19. McDaniel, P., and Mclaughlin, S. 2009. Security and privacy challenges in the smart grid. *IEEE Security & Privacy*, 7, 75-77.

20. Hayden, E. 2010. There is NO SMART in Smart Grid without Secure and Reliable Communications, Verizon Business Energy and Utility Solutions.

21. Victoria, T. and Pillitteri, Y. 2014. Guidelines for Smart Grid Cybersecurity. Technical Report, NIST. `https://www.nist.gov/publications/guidelines-smart-grid-cybersecurity` (acessed April 25, 2020).

22. Bari, A., Jiang, J., Saad, W., and Arunita, Jaekel. 2014. Challenges in the smart grid applications: An overview. *International Journal of Distributed Sensor Networks*, 10(2), 1-11.

23. Congressional Research Service. 2018. The Smart Grid: Status and Outlook. Technical Report, CRS. `https://crsreports.congress.gov/product/pdf/R/R45156` (accessed April 25, 2020).

24. Aggarwal, S., Chaudhary, R., Aujla, G.S., Jindal, A., Dua, A., and Kumar, N. 2018. Energychain: Enabling energy trading for smart homes using blockchains in smart grid ecosystem. In *1st ACM MobiHoc Workshop on Networking and Cybersecurity for Smart Cities* (pp. 1-6).

25. Vries, M. J. de. 2005. *Teaching About Technology. An Introduction to the Philosophy of Technology for Non-Philosophers.* Berlin: Springer-Verlag.

26. Zambroni de Souza, A.C., Bonatto, B.D., and Ribeiro, P.F. 2019. Emerging Smart Microgrid Power Systems: Philosophical Reflections. In *Microgrids Design and Implementation*, ed. A. Zambroni de Souza, M. Castilla, 505-528. Cham: Springer.

27. Verkerk, M., Hoogland, J., van der Stoep, J., and Vries, M. 2007. *Denken, ontwerpen, maken. Basisboek techniekfilosofie.* Amsterdam: Boom.

28. Kossiakoff, A., Sweet, W.N., Seymour, S.J., and Biemer, S.M. 2011. *Systems Engineering Principles and Practice: Second Edition*. Hoboken: Wiley-Interscience.
29. Dooyeweerd, H. 1969. *A New Critique of Theoretical Thought*. Phillipsburg: The Presbyterian and Reformed Publishing Company.
30. Vollenhoven, D.H.T. 2005. *Isagoogè Philosophiae (1967)*. Sioux Center: Dordt College Press.
31. Vasconsellos, P.H.N. 2019. Abordagem Filosófica e Visão Conceitual das Redes Elétricas Inteligentes. In *Integração de Renováveis e Redes Elétricas Inteligentes*, ed. A. Carlos Zambroni de Souza, 191-238. Rio de Janeiro: Interciência.
32. Ribeiro, P.F., Polinder, H., and Verkerk, M. 2012. Planning and designing smart grids: Philosophical considerations. *Technology and Society Magazine*, IEEE, 31, 34-43.
33. Jindal, A., Bhambhu, B. S., Singh, M., Kumar, N., and Naik, K. 2019. A heuristic-based appliance scheduling scheme for smart homes. *IEEE Transactions on Industrial Informatics*, 16(5), 3242-3255.
34. Baimel, D., Tapuchi, S., and Baimel, N. 2016. Smart grid communication technologies. *Journal of Power and Energy Engineering*, 04, 1-8.
35. Bayindir, R., Colak, I., Fulli, G., and Demirtas, K. (2016). Smart grid technologies and applications. *Renewable and Sustainable Energy Reviews*, 66, 499-516.
36. Yeung, P., and Jung, M. 2013. Improving Electric Reliability with Smart Meters. White Paper, Silver Spring Networks. http://61827ea6031ff2e7a9bc-6d82d48cc20ebafe1731d8b336d3561a.r18.cf2.rackcdn.com/uploaded/s/0e896733_1373576816_silverspring-whitepaper-improving-electric-reliability-smartmeters.pdf (accessed April 27, 2020).
37. Kantarci Melike, E., and Mouftah, H.T. 2013. Smart grid forensic science: Applications, challenges, and open issues. *Communications Magazine*, IEEE, 51, 68-74.
38. Althobaiti, A., Jindal, A., and Marnerides, A.K 2020. SCADA-agnostic Power Modelling for Distributed Renewable Energy Sources. In *IEEE 21st International Symposium on A World of Wireless, Mobile and Multimedia Networks (WoWMoM)* (pp. 379-384). IEEE.
39. Hossain, M.R., Oo, A.M.T., and Ali, A.B.M.S. 2013. Smart Grid, in *Opportunities, Developments, and Trends*, ed. A.B.M. Shawkat Ali, 23-44. London: Springer.
40. Gopakumar, P., Reddy, M.J.B., and Mohanta, D.K. 2017. Phasor measurement sensor based angular stability retention system for smart power grids with high penetration of microgrids. *IEEE Sensors Journal*, vol. 18(2), 764-772.
41. Segatto, M., Rocha, H., Lima Silva, J., Paiva, M., and Cruz, M. 2018. Telecommunication technologies for smart grids: Total cost optimization. *Advances in Renewable Energies and Power Technologies*, 14, 451-478.
42. Kato, D., Horii, H., and Kawahara, T. 2014. Next-generation SCADA/EMS designed for large penetration of renewable energy. *Hitachi Review*, 63(4), 151-155.
43. Acha, E., Fuerte-Esquivel, C.R., Ambriz-Perez, H., and Angeles-Camacho, C. 2004. *FACTS: Modelling and Simulation in Power Networks*. Chichester: John Wiley & Sons Ltd.
44. Yoldaş, Y., Önen, A., Muyeen, S.M., Vasilakos, A.V., and Alan, İ. 2017. Enhancing smart grid with microgrids: Challenges and opportunities. *Renewable and Sustainable Energy Reviews*, 72, 205-214.

45. JJ Justo, F Mwasilu, J Lee, and JW Jung 2013. AC-microgrids versus DC-microgrids with distributed energy resources: A review. *Renewable Sustainable Energies*, 24, 387–405.
46. Patrao, I., Figueres, E., Garcera, G., and Gonzalez-Medina, R. 2015. Microgrid architectures for low voltage distributed generation. *Renewable Sustainable Energies*, 43, 415-424.
47. Wollman, D., Greer, C., Prochaska, D., Boynton, P., Mazer, J., & Nguyen, Cuongo, et al. 2014. NIST Framework and Roadmap for Smart Grid Interoperability Standards. Technical Report, NIST. `https://nvlpubs.nist.gov/nistpubs/SpecialPublications/NIST.SP.1108r3.pdf` (acessed April 28, 2020).
48. Thomas, B., and Richard, De. 2011. IEEE Smart Grid Series of Standards IEEE 2030 (Interoperability) and IEEE 1547 (Interconnection) Status. Working Paper, NREL. `https://www.nrel.gov/docs/fy12osti/53028.pdf` (accessed April 28, 2020).
49. Jindal, A., Marnerides, A.K., Gouglidis, A., Mauthe, A., and Hutchison, D. 2019. Communication standards for distributed renewable energy sources integration in future electricity distribution networks. In *IEEE International Conference on Acoustics, Speech and Signal Processing (ICASSP)* (pp. 8390-8393). IEEE.
50. Philippe, B., and Hansman, R.J. 2008. Scalability of the Air Transportation System and Development of Multi-Airport Systems: A Worldwide Perspective. PhD. Thesis, Massachusetts Institute of Technology, Massachusetts.
51. Sigrist, L., May, K., Morch, A., Verboven, P., Vingerhoets, P., and Rouco, L. 2016. On scalability and replicability of smart grid projects—a case study. *Energies*, 9, 195.
52. Ma, S., Zhang, H., and Xing, X. 2018. Scalability for smart infrastructure system in smart grid: A survey. *Wireless Personal Communications*, 99, 161-184.
53. Rodriguez-Calvo, A., Cossent, R., and Frías, P. 2018. Scalability and replicability analysis of large-scale smart grid implementations: Approaches and proposals in Europe. *Renewable and Sustainable Energy Reviews*, 93, 1-15.
54. Aujla, G.S., Jindal, A., Kumar, N., and Singh, M. 2016. SDN-based data center energy management system using RES and electric vehicles. In *IEEE Global Communications Conference (GLOBECOM)* (pp. 1-6). IEEE.
55. Yinger, Bob. 2018. Integrated Grid Project (IGP). Planning Documentation, Southern California Edison, `https://www.sce.com/sites/default/files/inline-files/EPIC_IGPWinterSymposiumPresentation_0.pdf` (acessed May 10, 2020).
56. Public Service Company of New Mexico. 2028. Project Report, Department of Energy's Office of Electricity (DOE). `https://www.smartgrid.gov/project/public_service_company_new_mexico_pv_plus_battery_simultaneous_voltage_smoothing_and_peak` (accessed May 10, 2020)
57. Cheng, F., Willard, S., Hawkins, J., Arellano, B., Lavrova,O., and Mammoli A. 2012. Applying battery energy storage to enhance the benefits of photovoltaics. *IEEE Energytech Proceedings*, pp. 1-5.
58. Duke Energy Carolinas. 2011. PMU Deployment in the Carolinas with Communication System Modernization. Project Report, The Department of Energy's Office of Electricity (DOE). `https://www.smartgrid.gov/project/duke_energy_carolinas_llc_pmu_deployment_carolinas_communication_system_modernization` (accessed May 10, 2020).

# Section III

## *Internet of Energy (IoE)*

# 4 IoE: Solution for Smart Cities

*Ash Mohammad Abbas*
Aligarh Muslim University, India

## CONTENTS

## 4.1  INTRODUCTION

A smart city provides an adequate quality of life to its inhabitants using technological advancements. The data collected by different constituent departments and organizations of a smart city can be processed to improve the quality of services provided to its people. A smart city makes the life of its citizens easy and livable. A smart city should have necessary amenities for its inhabitants. For example, a smart city should have adequate water supply, energy supply, transport, schools, hospitals, etc. Out of these amenities, proper management of energy supply is crucial as it is required in all walks of life of the inhabitants of a smart city [5, 29, 30].

Application of networking to manage generation and consumption of energy for its effective utilization is called Internet of Energy (IoE). An IoE enables to enhance the utilization of energy and helps to mitigate the losses. It provides an infrastructure for communication of demands and supply of energy among producer and consumer sites. An IoE is perceived as a web enabled smart grid in [10]. An IoE enables to improve the quality of services provided to end users and optimize the revenues generated by energy providers. A description of preliminary concepts about IoE emphasizing its impact on electric mobility is provided in [63]. In [28], an IoE is perceived as an infrastructure that consists of smart sensor networks with schemes to manage the huge amount of data for smart grids [33]. Therein, different applications of smart sensor networks in case of smart grids are described. The role of IoE in different sections of smart power systems is studied in [58]. The supply and demand side sections of energy considered are renewable energy, bulk energy storage, power plant, and operation. Therein, the energy management in case of microgrids, vehicles to grid, and smart buildings is also considered.

Many researchers have focused on IoE from different perspectives [32]. A survey of IoT solutions for energy management in buildings that are part of a smart city is presented in [38]. A survey of research from the point of view of operation and planning of IoE is presented in [11]. It contains a description of an integrated energy system that consists of power systems and other related systems such as cooling & heating, natural gas, and traffic systems. Also, it contains a description of Combined Cooling & Heating Power (CCHP) system. A survey of IoE from the view point of its architecture and techniques is presented in [64]. Monitoring and control of energy requirements of buildings is carried out using Building Energy Management Systems (BEMS). There are buildings that consume as well as generate energy. The buildings where the effective energy consumption is zero or negligible are called net Zero Energy Buildings (nZEB). A survey of issues involved in IoE based BEMS is described in [22] taking into account nZEBs.

An overview of IoE based BEMS is presented in [49]. To solve the problems of data loss and overloading, a BEM is proposed. A survey of research describing the role of battery energy storage stations in IoE is carried out in [66]. It contains a description of the characteristics and key technologies for battery energy storage at power stations. A review of applications of Internet of Things (IoT) in case of energy sector with an emphasis on smart grids is presented in [48]. It contains a description of challenges in deploying IoT pertaining to security and their possible solutions using blockchain technology. Table 4.1 presents a summary of research surveys related to IoE.

An analysis of indices to evaluate energy efficiency in case of IoE is carried out in [59] based on the principles of thermodynamics. The indices evaluated are related to different sections of energy such as generation, transmission, conversion, storage, and consumption. These indices may help in the design and optimization of an IoE. The effect of different modes of energy utilization in regional IoEs is described in [74]. The modes of a regional IoE studied for utilizing energy are centralized, distributed, and mixed. Also, a comparison of the characteristics and applications of these modes is provided. A reference model for IoE called Open System Interconnection for Energy (OSI-E) is presented in [73]. It consists of the following five layers: access, transmission, distribution & control, strategy, and application. The accommodation of renewable energy in IoE is described in [72]. It contains a description of the key factors affecting the inclusion of renewable energy together with technologies such as optimal dispatch, integrated energy system and consumption of energy at customer sites. The use of IoE networks as a potential technology of future to manage energy with important lessons is described in [35].

**Table 4.1**

**A Summary of Research Surveys Related to Internet of Energy**

| Research Survey | Focus | Remarks |
|---|---|---|
| Khajenasiri et al. [38] | IoT solutions for energy management | nZEB |
| Cao et al. [11] | Operation & Planning of IoE | Integrated energy system |
| Wang et al. [64] | Architecture of IoE | Techniques for creation |
| Hannan et al. [22] | BEMS | Issues involved in IoE |
| Nguyen et al. [49] | BEMS | Data loss & overloading |
| Wei et al. [66] | Energy storage at power stations | Characteristics and key technologies |
| Motlagh et al. [48] | Application of IoT | Smart grids |

A lot of data is generated by different constituents of an IoE that need big data analytics and cloud computing for its processing [4, 40]. A computing platform for IoE using big data analytics and cloud computing is described in [20]. In case of Internet Data Centers (IDC), optimization of the energy consumption is a challenging task [2, 7]. The data centers that are geographically far apart may communicate about their energy demands using a data network and the loads of data centers can be transferred to one another for the purpose of load balancing. A scheme for managing energy in data centers based on workload optimization and CCHP is described in [42]. IDCs need to be equipped with energy storage devices for uninterrupted operation in case of emergencies. A description of the requirements and standards for storage of energy in data centers is provided in [3, 6, 36, 71]. It is discussed that energy storage devices should also draw energy from renewable energy sources.

There is a trend to reduce air pollution using renewable energy systems. An IoE makes the use of renewable energy systems easy and viable. It helps in reducing the air pollution using IoT enabled devices. Conversely, the use of IoE helps in improving the quality of air in a smart city. The impact of proliferation of IoE on the quality of air in China is evaluated in [41] using a spatial regression model in a smart city. An analysis of correlation between urban development and energy development is carried out in [13] to predict the development of clean and green energy in cities.

In a smart city, corporations that are responsible for power supply and consumer sites are connected through an IoE. There are many advantages of IoE in a smart city. For example, faults at consumer sites can be detected and remedial actions can be taken as soon as possible. The demands of energy for different constituents of a smart city can be estimated and a provision of a supply of an equivalent amount of energy can be made through an IoE. The demands in peak hours can be estimated and the pricing of energy can be determined to optimize the revenues generated by power supply corporations.

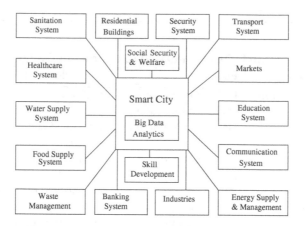

**Figure 4.1**  Different constituents of a smart city

The organization of this chapter is follows. Section 4.2, presents constituents of a smart city and the related research. Section 4.3 contains the need of IoE in smart cities. Section 4.4 consists of a description about the problems to be solved using IoE in smart cities. Section 4.5 contains a description about the operation of IoE in a smart city. Section 4.6 presents an integration of electrical vehicles to IoE. Section 4.7 contains a description of the infrastructure required for an IoE in a smart city. Section 4.8 contains examples of tools for IoE. The last section concludes the chapter.

## 4.2   CONSTITUENTS OF SMART CITIES

A smart city provides an adequate quality of life to its inhabitants and makes the life of its citizen easy and livable. Figure 4.1 shows the constituents of a smart city such as residential buildings, transport system, markets, education system, water supply system, food supply system, transport system, energy supply system, healthcare system, sanitation system, banking system, industries, security and policing, etc. All these systems exist to help the inhabitants of a city for making their life easy to live.

Many researchers have focused on smart cities from different perspectives. A description of the characteristics and components of a smart city is presented in [46]. There are six key components of a smart city namely people, economy, governance, technology, living, and environment. Therein, a seventh component called built environment is added to the list of key components of a smart city. It has been discussed that too much emphasis on Information Communication Technology (ICT) and business led models may have an adverse effect on interface dynamics of a smart city. A consideration of the built environment reveals that there should be a balance between the existing and arising infrastructures for a feasible development of a smart city.

Geospatial data may help in planning, designing, and operation of a smart city. A survey of research related to an improvement of services in smart cities using spatial computing is described in [67]. It is discussed that a Geographical Information

**Table 4.2**

**A Summary of Research Related to Participation of Citizens in a Smart City**

| Research | Focus | Governance Model |
|---|---|---|
| Ceballos et al. [12] | Satisfaction index | Citizen centric |
| Rhee et al. [57] | Development of smart cities | Collaboratory |
| Hayar et al. [23] | Transformation of a city into a smart city | Engagement of local people in development |
| Koutra et al. [39] | Multiscalar scheme for evolution and planning | Participatory |

System (GIS) data may help in analysis of alternative infrastructures of a smart city from different perspectives such as opportunities and risks. A case study of construction of cluster of smart cities using Technique of Order Preference Similarity to the Ideal Solution (TOPSIS) for three Chinese cities is presented in [43]. Note that TOPSIS is a method for analysis of multicriteria based decisions. It is useful for planning, designing, and improvement of services in a smart city where decisions are made using multiple criteria.

### 4.2.1 PARTICIPATION OF CITIZEN

Citizen are an important constituent of a smart city. The management of the facilities of a smart city should be carried out in consultation with its residents. A model for promoting the governance of a smart city by its citizen is described in [12]. It contains a description of metrics including an index related to the satisfaction of citizen of a smart city. The resources should be allocated taking into account the priorities of residents of a smart city. It can enhance the quality of life and the level of satisfaction among the residents of a smart city.

Challenges faced in IoT enabled smart cities are described in [57]. The development of smart cities can be motivated through collaboration among different constituents and stake holders. The IoT solutions that are cost-effective and scalable should be used in smart cities. A concept of an economical and socially sustainable smart city is presented in [23]. The engagement of citizen in the process of transformation of a city into a smart city can be an economical solution. To engage local people in the development of a city, one should promote innovations and provide incentives. A multiscalar scheme for the evolution and planning of a smart city is described in [39]. Therein, a description of the transformation of an urban area into a smart city in a participatory manner defining its objectives is presented. A summary of research related to participation of citizen in a smart city is provided in Table 4.2.

**Table 4.3**

**A Summary of Research Related to Residential Buildings in a Smart City**

| Research | Focus | Remarks |
| --- | --- | --- |
| Colistra et al. [14] | House infrastructure | Predictive healthcare |
| Ghosh et al. [21] | Architectural and engineering codes | Buildings and apartments |
| Ejaz et al. [18] | Optimization of energy consumption | Scheduling of IoT enabled devices |
| Kazmi et al. [37] | Smart home | Home energy management system |
| Bampoulas et al. [9] | On demand response | Machine learning algorithms |

## 4.2.2 RESIDENTIAL BUILDINGS

The evolution of the architecture of smart cities is presented in [14] from the point of view of development of housing infrastructure with a multiple family prototype. The data gathered from the housing infrastructure can be used for predictive healthcare in smart cities. Homes and apartments are the basic building blocks in a smart city. There should be architectural and engineering codes for buildings in a smart city. The areas in a smart city where such codes are required are explored in [21]. The use of IoT for energy efficient operation in smart cities is described in [18]. A framework for optimizing energy consumption and scheduling of IoT enabled devices is presented with an emphasis on energy harvesting in smart cities. Note that energy harvesting extends the lifetime of low power devices. It contains case studies of energy-efficient operation of smart homes and wireless power transfer for IoT enabled devices in smart cities.

Optimization of heuristic and metaheuristic algorithms for a balance among the demand and supply of energy in case of IoT enabled smart homes is presented in [31, 37]. A smart home is supposed to have a home energy management (HEM) system that is responsible for managing the demand side energy. An HEM schedules the appliances according to the supply of energy, automatically controls them and shifts their load from peak to off-peak hours. Therein, the performance of metaheuristic algorithms such as harmony search and differential evolution is evaluated. The effect of integration of renewable energy sources has also been evaluated. The problem of the use of renewable energy sources in peak hours is formulated using a knapsack problem. The use of machine learning algorithms in energy systems for residential buildings is described in [9]. Therein, the major focus is on demand response that provides flexibility of energy usage and affects the satisfaction of customers. A summary of research related to residential buildings in a smart city is presented in Table 4.3.

**Table 4.4**
**A Summary of Research Related to Street Lights in a Smart City**

| Research | Focus | Remarks |
|---|---|---|
| Ozadowicz et al. [51] | Energy consumption | Remote monitoring |
| Sudarmono et al. [61] | Energy consumption | Replace HPS with LED |
| Rabaza et al. [55] | Maps of street lights | GIS |
| Yang et al. [68] | Management of street lights | Docker |
| Pandharipande et al. [52] | Connected street lights | Data Models |

### 4.2.3  STREET LIGHTS

A major component of energy consumption in cities is street lights. A method for reducing energy consumed by street lights in a smart city is proposed in [51]. It contains a scheme for controlling the lamps of street lights together with a function for remote monitoring. An impact of Light Emitting Diode (LED) lights in streets on energy efficiency of Jakarta province of Indonesia is evaluated in [61]. To reduce the energy consumption, High Pressure Sodium (HPS) armatures were replaced with LED armatures. A method for measuring the level of lighting in different portions of a smart city through aerial images is described in [55]. The measurements of luminous intensity, energy consumption, and installation devices are combined with a Geographical Information System (GIS). The maps of street lights can be generated to extract the information about lighting system in a smart city. It enables to initiate remedial actions to correct the problems and make it more efficient and viable.

A system for managing street lights is proposed in [68]. It consists of a cloud management platform, edge devices, and a light control system. A user may obtain information about the street poles in real time. Additionally, an architecture that consists of a container-based virtualization technique for providing cloud and edge services is also proposed. A secure communication between edge devices and the cloud is provided through Secure Shell (SSH). An architecture for street lights in a smart city is proposed in [52]. It consists of models of different types of data for the design of Application Programming Interfaces (API). Connected street lights can be helpful for applications such as autonomous driving, networking, and environmental monitoring. A summary of research related to street lights is provided in Table 4.4.

### 4.3  NEED OF IOE IN SMART CITIES

The consumption of energy is increasing all over the world due to several factors. Some of them are as follows.

*Increasing population*: An increase in the population implies an increase in the number of homes that requires an increase in the energy consumed in heating in cold areas and cooling and air conditioning in warm areas. Also, an increased number of homes requires an increase in lights, fans, cooking, refrigeration, vehicles, street lights, roads, transports, etc., causing an increased amount of energy.

*Increase in industries*: Technological advancements and increase in the population have increased the needs of people. To fulfill the needs of people requires an increased number of industries. It gives rise to an increase in the energy consumption.

*Increase in electrical vehicles*: There is a significant increase in the number of electrical vehicles, such as metro trains, chargeable battery operated electric cars, auto rickshaws, etc. The batteries of electrical vehicles are often charged using electricity from the smart grid.

*Increase in living standards*: With technological advancements, the living standards of people have increased in almost all parts of the world. More and more devices, equipment, vehicles, etc. have become part of the daily lives of people. As a result, the consumption of energy has increased throughout the world.

*Changing life style of people*: The life style of people has changed. Previously when electric energy was not available in large quantities, people used to sleep early in the night and wake up early in the morning. With the availability of electric energy in every walk of life, the sleeping habits, work hours, marketing, dining, and entertainment times of people have changed. In many parts of the worlds, markets, restaurants, concert halls, hotels, and hospitals are open till late nights giving rise to an increase in the consumption of energy.

If the demands and supplies of energy are not communicated, it may happen that in some areas the demands of energy may exceed the supply of energy and in other areas supply of energy may exceed the demands of energy. It may cause power cuts in areas where supply of energy falls short of demands depriving people of their energy needs. Also, excess supplies and lack of supplies decreases the utilization of energy and may cause a decrease in the revenues collected by the energy generating organizations. In other words, there is a need for demands and supplies of energy to be communicated over a network called an IoE.

## 4.4   PROBLEMS TO BE SOLVED USING IOE

In a smart city, where the services need to be automated and should be delivered in the best possible manner, the role of IoE becomes very important. An IoE can be used to provide better services to end users and optimize the revenues of the organizations generating the energy. The problems that an IoE may solve in a smart city are as follows.

*Power outages*: If the demands and supplies of energy are not communicated, there can be power failures or black outs or unnecessary power cuts.

An IoE can fill the gap between the demand and the supply of energy. The faults in different entities can be detected early and measures can be taken before a power failure that may wreak a havoc.

*Handling of peak hours*: An IoE enables to determine the peak hour loads and enables to send control signals to consumers to reduce their loads during peak hours. Also, it enables to send control signals to power stations to increase the production of energy, if possible, during peak hours.

*Reduction in pollution*: Renewable sources of energy such as the solar power should be encouraged over non-renewable sources of energy such as fossil fuels for reducing the air pollution. For example, for transportation in a smart city, the use of electrical vehicles should be encouraged to reduce air pollution. In other words, the IoE has a potential to provide clean energy and to enable the smart cities to be green or energy-efficient.

*Optimization of revenues*: An IoE helps in optimizing revenues of energy producing corporations that are generated from the customers. Rates of energy can be fixed based on several factors, such as the utilization during the peak hours, the amount of load, season, etc.

*Service quality*: The IoE enables to make decisions based on analysis of a huge amount of data collected from customer sites, substations, energy generating and distributing sites. Necessary actions are taken as soon as a fault occurs and hence it enables to improve the quality of services provided to end users.

In a smart city, there can be many types of energies coming from different sources, and there can be many types of end users with different needs. In what follows, we describe the operation of an IoE.

## 4.5  OPERATION OF IOE

The electricity generated from different sources is added to the grid. A grid with computational and communication capabilities at the stations is often called a smart grid. An IoE is a network of end user devices and energy providers that are equipped with sensors to collect data and forward it using a communication network to control stations for further processing and decision making. A network of appliances such as refrigerators, washing machines, dish washers, television, chandeliers, microwave ovens, electric ovens, vacuum cleaners, water coolers, air conditioners, room heaters, etc., connected to the Internet is called an Internet of Things (IoT). In an IoE, all end user devices including energy meters are IoT enabled.

In a smart city, the supply of energy comes from different sources through a smart grid. The sources can be nuclear reactors, thermal power stations, hydro power stations, solar parks, etc. All these energy providers might be at remote locations. At the consumer side, there can be industries, commercial places, academic institutions, municipal corporations, departments and offices, residential buildings and homes, etc. Some of the consumers might also be producing some forms of energy e.g. solar energy through roof top solar cells that needs to be added to the smart grid. The

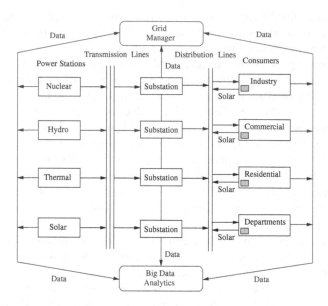

**Figure 4.2**   Operation of IoE in a smart city

solar panels are mounted on the roofs of the buildings within the smart city reducing the dependence of the energy requirements of the city on the sources other than the resources of the city.

Figure 4.2 shows the operation of an IoE in a smart city. There are different types of major sources of energy or power stations such as nuclear, hydro, thermal, and solar power stations feeding a smart grid. There are different categories of consumers of energy in a smart city, for example, industries, big commercial houses, residential buildings, departments, educational institutes. Some of the consumers may also produce solar energy using roof top solar panels and the energy generated might be fed to the smart grid. The major steps from the generation till the consumption of power are as follows: (i) generation, (ii) transmission, (iii) distribution, and (iv) consumption.

The power is generated at power plants at relatively low voltage ranging from 2.3 kV to 30 kV that depends on the size of power generating units. It is fed to a step up transformer that converts it to high voltage power that is carried to distant places using transmission lines. The typical voltage levels for transmission are from 115 kV to 765 kV for AC depending on the transmission system and the region. Sometimes, lower voltage levels such as 66 kV and 33 kV are used for long distance transmission of relatively low power. For long distances, the transmission of power using Over Head Power (OHP) cables is preferred over under-ground cables due to ease of operation and maintenance. The high voltage transmission reduces the loss of energy due to the resistance that depends on the length of the transmission line, conductor, and the amount of current. Specifically, the loss of energy during transmission is $I^2R$ where $I$ is the current and $R$ is the resistance. Making the voltage high implies decreasing the current and consequently reducing the losses during transmission. Further, transmission of power at much higher voltage levels such as greater than or equal to 2000 kV is not recommended because the loss of energy due to *corona effect* overcomes the resistive loss.

From the transmission lines, the power is fed to substations responsible for distributing it in a region. The distributing stations use step down transformers to decrease the voltage of the power. For distribution in a city or a region, the distribution lines use the voltage of 33 kV or less. From the substation, there are distribution lines that carry the power toward different types of customers in a locality. There are often pole mounted step down transformers that further decrease the voltage so that it can be used at the low voltage consumer sites or houses. For consumption in residential homes and buildings, the voltage levels are often 240 V.

The power stations, substations, and consumers generate a lot of data. This data is gathered by sensors and sent toward a computing station housing high computing servers and software. The data received by the computing station is processed using big data analytics techniques. The processed information is sent toward the grid manager that is responsible for sending control signals to different entities. Note that there can be more than one computing stations and grid managers in a smart city depending on the volume of data to be processed. All these computing stations and grid managers may communicate with one another and cooperate in performing their functions. Each computing element is responsible to process data gathered by sensors in its region. Similarly, each grid manager is responsible to take decisions and send control signals to entities in its region.

A description of architectures and communication technologies for smart grids and their transition to IoE is provided in [34]. A software defined network (SDN) framework for communication in an IoE is proposed in [44]. The framework is hierarchical in nature and contains clusters of microgrids connected to a global grid. An architecture of an IoE containing a combination of energy management and control systems is proposed in [65]. The architecture contains four layers and the functions of each layer are analyzed. The architecture is adaptable, flexible, and comprehensive. Prospects of research related to virtual microgrids using IoE are described in [69]. It includes a description of the architecture, operation, and key mechanisms related to virtual microgrids or virtual power plants. An architecture for an IoE with an emphasis on communication in smart grid is presented in [56]. An H-infinity based min-max filter is presented to estimate the states of a microgrid. A fog-based architecture for IoE to manage the transactive energy is proposed in [47]. The architecture consists of home gateways to collect data about energy consumption, fog nodes at the network edge to act as retail energy servers, and cloud servers to provide high-speed computing and bulk data storage. A summary of research related to architectures of IoE is presented in Table 4.5.

**Table 4.5**
**A Summary of Research Related to Architectures of IoE**

| Architecture | Focus | Remarks |
|---|---|---|
| Kabalci et al. [34] | Architecture and communication technologies | Transition of smart grid to IoE |
| Lu et al. [44] | SDN-enabled framework | Hierarchical |
| Wang et al. [65] | Combination of energy management & control systems | Four-layered architecture |
| Yang et al. [69] | Architecture, operation, & key mechanisms | Virtual microgrids |
| Rana et al. [56] | Communications in smart grid | H-infinity-based min-max filter |
| Moghaddam et al. [47] | Fog-based architecture | Management of transactive energy |

## 4.6 INTEGRATION OF ELECTRICAL VEHICLES TO IOE

Petroleum is often used for transport, however, in oil exporting countries, it is also used to generate electricity. A lot of research is going on to use energy in transportation other than that provided by petroleum products so as to reduce the dependence on fossil fuels. For transport, a lot of research is being carried out for developing Electrical Vehicles (EV) such as Battery Electric Vehicles (BEV) and Hybrid Electrical Vehicles (HEV). In contrast to BEVs that are fully battery chargeable vehicles, HEVs have options for their batteries to be charged using both electricity and gasoline.

A method for optimization of energy used by EVs in a smart city is described in [1, 8]. Vehicles and buildings communicate for sharing information about status of vehicle and road conditions. The information about the energy consumption along a specific road segment can be shared by EVs to a data base and can be used to manage energies of EVs planning to travel along the road segment. Artificial Intelligence (AI) techniques such as Support Vector Machine (SVM) can be used to classify vehicles based on recommendations of buildings about the positions of the vehicles.

It has been pointed out in [26] that EVs can be charged using vehicle-to-grid (V2G) infrastructure. However, a large number of vehicles should not be charged simultaneously to avoid stress on the grid. The batteries of vehicles can be scheduled for charging in parking spaces. There can be both consumption of energy while charging using the grid and generation of energy while it is moving on the roads. The energy generated can be transferred from vehicles to the grid. In other words, there should be a smart charging infrastructure for charging vehicles at homes and places of work.

**Table 4.6**

**A Summary of Research Related to the Integration of EVs and IoE**

| Research Survey | Focus | Remarks |
|---|---|---|
| Aymen et al. [8] | Optimization of energy | Smart city |
| IRENA. [26] | Smart charging | V2G infrastructure |
| Mahmud et al. [45] | Integration of EVs & IoE | Architecture for management of energy |
| Vaidya et al. [62] | CAEV | Use cases |
| Su et al. [60] | Secure charging | Contract-based blockchain |
| Yi et al. [70] | Energy network | Optimal placement of charging stations |

EVs may act as mobile power generators whose kinetic energy can be converted to electrical energy. EVs may connect to an IoE where the flow of information and energy is bidirectional. However, an integration of EVs and an IoE requires many issues to be addressed, such as charging schemes, battery technologies, transfer of energy, communication, and standards. A review of challenges, issues and possible solutions related to EV technologies, charging schemes, and smart charging is presented in [45]. Also, an architecture for managing the energy of distributed EVs using an IoE is presented. A description of an implementation of Connected and Autonomous Electric Vehicles (CAEV) and related use cases is provided in [62].

Integration of EVs and distributed energy sources to the smart grid can be carried out using an IoE. However, in a smart community (SC) environment, the charging of EVs need to be secure. A contract-based blockchain for secure charging of EVs is presented in [60]. It contains an energy blockchain, a fault tolerant algorithm based consensus, and a mechanism to allocate energy. A model of network for transmission, storage, and distribution of energy generated by EVs is presented in [70]. For an optimal placement of charging stations in the energy internet, solutions using greedy heuristic and diffusion algorithms are described. A summary of research related to an integration of EVs to an IoE is provided in Table 4.6.

## 4.7 INFRASTRUCTURE REQUIRED FOR IOE

In a smart city, collection of data about consumption of energy by homes, industrial units, offices, departments, hotels, academic units, etc., should be automated. Also, collection of data about generation of energy by all these units should also be automated to compute the effective energy requirements of a smart city. An analysis of energy consumption such as maximum, minimum, and average energy consumption during a day, a month, and a year should be carried out. Such an analysis can help in estimating the requirements of energy and detecting anomalous behavior or possibility of a fault so that appropriate actions can be taken in a timely manner.

Similarly, an analysis of energy generated by resources within the smart city should be carried out. An analysis of losses of energy in distribution should also be carried out. The pricing of energy can be based on the processing of the data collected and its analysis. For that purpose, the energy meters installed at homes, buildings, departments, and offices in a smart city should have sensors to collect data and forward it to control stations for further processing. Alternatively, the old energy meters should be phased out or upgraded to add the desired functionality.

Industries should be encouraged to use energy-efficient machines for producing their items. The companies making electrical and electronic devices should be encouraged to produce energy-efficient and cost-effective devices and equipment such as home appliances, fans, geysers, refrigerators, washing machines, dish washers, cooking ovens, computers, mobile phones, space heaters, and air conditioners.

All lights on streets and roads should draw energy from solar panels. However, after the depletion of solar energies, they should switch to draw the energy from the smart grid. Old lights on streets and roads should be phased out and should be replaced with the latest cost-effective and energy-efficient lights. Consumers should be encouraged to produce and use energy from renewable resources such as solar power. Incentives should be provided for production of solar energy at consumer premises and its transfer to the smart grid.

Battery chargeable electric vehicles such as bikes, auto rickshaws, cars, and buses should be proliferated and encouraged to be used. In each smart city, there should be proper public transport system and its use should be encouraged. At peak hours when the demand of energy is high, either the generation of power should be increased or consumers should reduce their loads. Alternatively, incentives should be provided to consumers to use appliances requiring more power in non-peak hours. The incentives can be in various forms e.g. in terms of reduced rates in off-peak hours.

## 4.8   IOE TOOLS

There are many types of tools available for IoE in a smart city including power consumption monitoring, fault detection, and control, etc. There are the following tools that support IoE.

> *General Electric Current*: The General Electric (GE) provides a wide range of products for optimization and management of energy [15] in automated industries, smart banking, smart cities, offices, and grocery stores. It provides LED lights in terms of indoor & outdoor fixtures, and control & sensors. The indoor fixtures include disinfection lights, recessed lights, architectural systems, industrial lights, low & high bay lights, etc. The outdoor fixtures include lights for roads & streets, decorative post tops, parking spaces, etc. In addition, it provides lights for different applications such as horticulture, hazardous locations, railways and traffic control, etc.
>
> *Philips Hue*: The Philips Hue [24] provides a range of products including essentials, such as starter kits, bridge, and accessories; lamps & lights, such as bulbs, lamps, fixtures; and new products, such as Bluetooth lights, outdoor light strips, etc. It provides smart lights for residential homes in a smart city.

These include smart lights for living room, bed room, kitchen, parking, etc. These lights are often voice controlled.

*Neo Carbon Energy*: Neo Carbon Energy [19] is a research project for renewable energy in cooperation with Lappenranta University of Technology (LUT), VTT Technical Research Center of Finland, and Finland Future Research Center (FFRC). It provides visualization tool for simulation of a renewable energy system.

In addition to them, many companies are offering IoT tools that can be helpful in IoE.

*Smart Camera and Thermostat*: Ecobee [16] provides smart camera for home monitoring, smart thermostats, and smart sensors for energy efficient operations of doors and windows. These smart devices have voice control. Ecobee also provides air filters and solutions for smart buildings.

*EdgeX*: WebNMS provides a set of IoT tools including EdgeX [17] that can be helpful in IoE for energy management, smart metering, smart lighting, etc. EdgeX provides a light weight, secure, reliable and robust protocol suite. It is easy to configure and vendor neutral.

*Gateway for Energy Management*: Intel provides a gateway [53] based on a System-on-Chip (SoC) called Quark$^{TM}$ for management of energy. It enables to connect electrical devices to the cloud remotely and to find methods for reducing the consumption of energy.

*Street Lights for Smart Cities*: GridComm provides a solution for networking of street lights in a smart city [54] using an IoT platform from Intel. Tvilight CityManager [25] is a scalable and secure software containing open APIs for the management of street lights in a smart city. Also, it provides an IoT cloud platform called DigiHub, lighting control and services. DigiHub gathers data from devices and gateways, analyzes and provides the data to CityManager and other open APIs. Itron provides IoT solutions for intelligent street lighting [27] in smart cities.

Many companies involved in the energy sector are focusing on IoT enabled energy management [50] including Duke Energy, Pacific Gas & Electric Company, EDF, National Grid, Nissan, Brooklyn Microgrid, EON, RWE, Hive, etc.

## 4.9 CONCLUSION

In this chapter, we presented IoE as a solution for energy requirements in smart cities. We described the need of IoE in a smart city and the problems that can be solved using IoE. We described the operation of an IoE in a smart city. Also, we presented the infrastructure required by an IoE in a smart city. We described tools that may help in IoE for energy efficient lighting, energy management, monitoring and control of energy in a smart city. There are lots of opportunities for professionals pertaining to the establishment of IoE in smart cities. A lot of research is being carried out and is required in this area for the improvement of services provided to inhabitants of a smart city using IoE.

## REFERENCES

1. Gagangeet Singh Aujla, Anish Jindal, and Neeraj Kumar. Evaas: Electric vehicle-as-a-service for energy trading in sdn-enabled smart transportation system. *Computer Networks*, 143:247–262, 2018.
2. Gagangeet Singh Aujla and Neeraj Kumar. Mensus: An efficient scheme for energy management with sustainability of cloud data centers in edge–cloud environment. *Future Generation Computer Systems*, 86:1279–1300, 2018.
3. Gagangeet Singh Aujla and Neeraj Kumar. Sdn-based energy management scheme for sustainability of data centers: An analysis on renewable energy sources and electric vehicles participation. *Journal of Parallel and Distributed Computing*, 117:228–245, 2018.
4. Gagangeet Singh Aujla, Neeraj Kumar, Mukesh Singh, and Albert Y Zomaya. Energy trading with dynamic pricing for electric vehicles in a smart city environment. *Journal of Parallel and Distributed Computing*, 127:169–183, 2019.
5. Gagangeet Singh Aujla, Maninderpal Singh, Arnab Bose, Neeraj Kumar, Guangjie Han, and Rajkumar Buyya. Blocksdn: Blockchain-as-a-service for software defined networking in smart city applications. *IEEE Network*, 34(2):83–91, 2020.
6. Gagangeet Singh Aujla, Mukesh Singh, Neeraj Kumar, and Albert Zomaya. Stackelberg game for energy-aware resource allocation to sustain data centers using res. *IEEE Transactions on Cloud Computing*, 7(4), 1109-1123, 2017.
7. Gagangeet Singh Aujla, Neeraj Kumar, Sahil Garg, Kuljeet Kaur, and Rajiv Ranjan. Edcsus: Sustainable edge data centers as a service in sdn-enabled vehicular environment. *IEEE Transactions on Sustainable Computing*, DOI: 10.1109/TSUSC.2019.2907110 2019.
8. Flah Aymen and Chokri Mahmoudi. A novel energy optimization approach for electrical vehicles in a smart city. *Energies*, 2019 (12):1–22, 2019.
9. Adamantios Bampoulas, Mohammad Safari, Fabiano Pallonetto, Eleni Mangina, and Donal P. Finn. Self-learning control algorithms for energy systems integration in the residential building sector. In *Proceedings of IEEE 5th World Forum on Internet of Things (WF-IoT)*, pages 815–818, Limerick, Ireland, April 2006. IEEE.
10. Nicola Bui, Angelo P. Castellani, Paolo Casari, and Michele Zorzi. The Internet of Energy: A web-enabled smart grid system. *IEEE Network*, 26(4):39–45, 2019.
11. Yijia Cao, Qiang Li, Yi Tan, Yong Li, Yuanyang Chen, Xia Shao, and Yao Zou. A comprehensive review of energy internet: Basic concept, operation and planning methods, and research prospects. *Journal of Modern Power Systems and Clean Energy*, 6(3):399–411, 2018.
12. Gonzalo R. Ceballos and Victor M. Larios. A model to promote citizen driven government in a smart city: Use case at gdl smart city. In *Proceedings of IEEE International Smart Cities Conference (ISC2)*, pages 1–6, Trento, Italy, September 2016. IEEE.
13. Yufeng Chai, Cuifen Bai, Chen Zhang, Xinyang Han, Xianzhong Dai, and Xiaoling Jin. Energy internet system research based on development trend of urban energy system. In *Proceedings of IEEE 3rd Conference on Energy Internet and Energy System Integration (EI2)*, pages 604–609, Changsha, China, November 2019. IEEE.

14. Joe Colistra. The evolving architecture of smart cities. In *Proceedings of IEEE International Smart Cities Conference (ISC2)*, pages 1–8, Kansas City, MO, USA, September 2018. IEEE.

15. GE Current. Reveal the potential of intelligent environments. `https://www.gecurrent.com/`, 2020. [Online; accessed 29-April-2020].

16. Ecobee. Your home, as you imagine it. `https://www.ecobee.com/`, 2020. [Online; accessed 29-April-2020].

17. WebNMS EdgeX. Seamless connectivity to power your IoT application. `https://www.webnms.com/iot/edgex-agent.html`, 2020. [Online; accessed 29-April-2020].

18. Waleed Ejaz, Muhammad Naeem, Adnan Shahid, Alagan Anpalagan, and Minho Jo. Efficient energy management for the internet of things in smart cities. *IEEE Communications Magazine*, 55(1):84–91, 2017.

19. Neo Carbon Energy. Trust in renewable, emission free future now available. `http://www.neocarbonenergy.fi/`, 2020. [Online; accessed 1-May-2020].

20. Rui Fu, Feng Gao, Rong Zeng, Jun Hu, Yi Luo, and Lu Qu. Big data and cloud computing platform for energy internet. In *Proceedings of China International Electrical and Energy Conference (CIEEC)*, pages 681–686, Beijing, China, October 2017. IEEE.

21. Sameek Ghosh. Smart homes: Architectural and engineering design imperatives for smart city building codes. In *Proceedings of IEEE Conference on Technologies for Smart-City Energy Security and Power (ICSESP)*, pages 1–4, Bhubaneswar, India, March 2018. IEEE.

22. Mahammad A. Hannan, Mohammad Faisal, Pin Jern Ker, Looe Hui Mun, Khadija Parvin, Teuku Meurah Indira Mahlia, and Frede Blaabjerg. A review of internet of energy based building energy management systems: Issues and recommendations. *IEEE Access*, 6:38997–39014, July 2018.

23. Aawatif Hayar and Gilles Betis. Frugal social sustainable collaborative smart city casablanca paving the way towards building new concept for "future smart cities by and for all". In *Proceedings of IEEE Conference on Sensors Networks Smart and Emerging Technologies (SENSET)*, pages 1–4, Beirut, Lebanon, September 2017. IEEE.

24. Philips Hue. Smart home lighting made brilliant. `https://www2.meethue.com/`, 2020. [Online; accessed 29-April-2020].

25. Tvilight: Empowering Intelligence. City Manager, Smart Street Lighting for Smart(er) Cities. `https://www.tvilight.com/wp-content/uploads/2019/03/Brochure-Intelligent-Smart-City-Lighting-Control-Sensor-CMS-Light-Management-EN.pdf`, 2020. [Online; accessed 21-July-2020].

26. International Renewable Energy Agency (IRENA). Smart Charging: Parked EV batteries can save billions in grid balancing. `https://energypost.eu/smart-charging-parked-ev-batteries-can-save-billions-in-grid-\cing/` 2020. [Online; accessed 23-July-2020].

27. Itron. Intelligent Street Lighting: Illuminating IoT. `https://www.itron.com/na/solutions/what-we-enable/smart-cities/intelligent-streetlights`, 2020. [Online; accessed 20-July-2020].

28. Manar Jaradat, Moath Jarrah, Abdelkader Bousselham, Yaser Jaraweh, and Mohammad Ali Ayyoub. The Internet of Energy: Smart sensor networks and big data management for smart grid. *Procedia Computer Science*, 6:592–597, 2015.

29. Anish Jindal. *Data Analytics of Smart Grid Environment for Efficient Management of Demand Response*. PhD thesis, Thapar Institute of Engineering and Technology Patiala, 2018.

30. Anish Jindal, Gagangeet Singh Aujla, Neeraj Kumar, Radu Prodan, and Mohammad S Obaidat. Drums: Demand response management in a smart city using deep learning

and svr. In *2018 IEEE Global Communications Conference (GLOBECOM)*, pages 1–6. IEEE, 2018.

31. Anish Jindal, Bharat Singh Bhambhu, Mukesh Singh, Neeraj Kumar, and Kshirasagar Naik. A heuristic-based appliance scheduling scheme for smart homes. *IEEE Transactions on Industrial Informatics*, 16(5):3242–3255, 2019.

32. Anish Jindal, Neeraj Kumar, and Mukesh Singh. Internet of energy-based demand response management scheme for smart homes and phevs using svm. *Future Generation Computer Systems*, 108:1058–1068, 2020.

33. Anish Jindal, Neeraj Kumar, and Mukesh Singh. A unified framework for big data acquisition, storage, and analytics for demand response management in smart cities. *Future Generation Computer Systems*, 108:921–934, 2020.

34. Ersan Kabalci and Yasin Kabalci. *Principal component neural networks: Theory and applications*. Academic Press, Cambridge, Massachusetts, USA, first edition, 2019.

35. Shivkumar Kalyanaraman. Back to the future: Lessons for internet of energy networks. *IEEE Internet Computing*, 20(1):60–65, January 2016.

36. Devinder Kaur, Gagangeet Singh Aujla, Neeraj Kumar, Albert Y Zomaya, Charith Perera, and Rajiv Ranjan. Tensor-based big data management scheme for dimensionality reduction problem in smart grid systems: Sdn perspective. *IEEE Transactions on Knowledge and Data Engineering*, 30(10):1985–1998, 2018.

37. Saqib Kazmi, Nadeem Javaid, Muhammad Junaid Mughal, Mariam Akbar, Syed Hassan Ahmed, and Nabil Alrajeh. Towards optimization of metaheuristic algorithms for IoT enabled smart homes targeting balanced demand and supply of energy. *IEEE Internet Computing*, 7:24267–24281, October 2017.

38. Iman Khajenasiri, Abouzar Estabsari, Marian Verhelst, and Georges Gielen. A review on internet of things solutions for intelligent energy control in buildings for smart city applications. *Energy Procedia*, 111:770–779, March 2017.

39. Sesil Koutra, Vincent Becue, and Christos S. Ioakimidis. A multiscalar approach for smart city planning. In *Proceedings of IEEE International Smart Cities Conference (ISC2)*, pages 1–7, Trento, Italy, September 2016. IEEE.

40. Neeraj Kumar, Tanya Dhand, Anish Jindal, Gagangeet Singh Aujla, Haotong Cao, and Longxiang Yang. An edge-fog computing framework for cloud of things in vehicle to grid environment. In *2020 IEEE 21st International Symposium on "A World of Wireless, Mobile and Multimedia Networks" (WoWMoM)*, pages 354–359. IEEE, 2020.

41. Lei Li, Yilin Zheng, Shiming Zheng, and Huimin Ke. The new smart city programme: Evaluating the effect of the internet of energy on air quality in China. *Science of The Total Environment*, 714, April 2020.

42. Qiang Liu, Songsong Chen, Min Chen, and Ciwei Gao. Energy management for internet data centers considering the coordinating optimization of workload and cchp system. In *Proceedings of 2nd IEEE Conference on Energy Internet and Energy System Integration (EI2)*, pages 1–5, Beijing, China, October 2018. IEEE.

43. Yongbo Liu. The study on smart city construction assessment based on topsis "the beijing-tianjin-tangshan city clusters" as the case. In *Proceedings of IEEE International Conference on Smart City and Systems Engineering (ICSCSE)*, pages 321–325, Hunan, China, November 2016. IEEE.

44. Zhaoming Lu, Chunlei Sun, Jinqian Cheng, Yang Li, Yong Li, and Xiangming Wen. SDN-enabled communication network framework for energy internet. *Journal of Computer Networks and Communications*, 2017:1–13, June 2017.

45. Khizir Mahmud, Graham E. Town, Sayidul Morsalin, and M.J. Hossain. Integration of electric vehicles and management in the internet of energy. *Renewable and Sustainable Energy Reviews*, 82(3):4179–4203, February 2018.

46. Emile Mardacany. Smart cities characteristics: Importance of built environments components. In *IET Conference on Future Intelligent Cities*, pages 1–6, London, UK, December 2014. IET.

47. Mohammad Hossein Yaghmaee Moghaddam and Alberto Leon-Garcia. A fog-based internet of energy architecture for transactive energy management systems. *IEEE Internet of Things Journal*, 5(2):1055–1069, April 2018.

48. Naser Hossein Motlagh, Mahsa Mohammadrezaei, Julian Hunt, and Behnam Zakeri. Internet of Things (IoT) and the energy sector. *Energies*, 2020(13):1–27, January 2020.

49. Van Thang Nguyen, Thanh Luan Vu, Nam Tuan Le, and Yeong Min Jang. An overview of internet of energy (ioe) based building energy management system. In *Proceedings of IEEE International Conference on Information and Communication Technology Convergence (ICTC)*, pages 852–855, Jeju Island, Korea, October 2018. IEEE.

50. Internet of Business. 10 real-life examples of IoT powering the future of energy. https://internetofbusiness.com/10-examples-showcasing-iot-energy/, 2020. [Online; accessed 29-April-2020].

51. Andrzej Ozadowicz and Jakub Grela. Energy saving in the street lighting control system: A new approach based on the EN-15232 standard. *Energy Efficiency*, 10(2):563–567, June 2017.

52. Ashish Pandharipande and Paul Thijssen. Connected street lighting infrastructure for smart city applications. *IEEE Internet of Things Magazine*, 2(2):32–36, June 2019.

53. Intel White Paper. Energy management framework greatly reduces effort for solution providers. https://www.intel.in/content/www/in/en/energy/quark-soc-energy-management-gateway-paper.html, 2020. [Online; accessed 30-April-2020].

54. Intel IoT Platform. Street lighting, bringing light to the internet of things. https://www.intel.in/content/www/in/en/internet-of-things/solution-briefs/gridcomm-solution-brief.html, 2020. [Online; accessed 20-April-2020].

55. Ovidio Rabaza, Evaristo Molero-Mesa, Fernando Aznar-Dols, and Daniel Gmez-Lorente. Experimental study of the levels of street lighting using aerial imagery and energy efficiency calculation. *Sustainability*, 10(12), p.4365, November 2018.

56. Md Masud Rana. Architecture of the internet of energy network: An application to smart grid communications. *IEEE Access*, 5:4703–4710, 2017.

57. Sokwoo Rhee. Catalyzing the internet of things and smart cities: Global cityteams challenge. In *Proceedings of IEEE 1st International Workshop on Science of Smart City Operations and Platforms Engineering (SCOPE) in partnership with Global City Teams Challenge (GCTC)*, pages 1–4, Vienna, Austria, April 2016. IEEE.

58. Hossein Shahinzadeh, Jalal Moradi, Gevork B. Gharehpetian, Hamed Nafisi, and Mehrdad Abedi. Internet of energy (ioe) in smart power systems. In *Proceedings of 5th IEEE Conference on Knowledge Based Engineering and Innovation (KEBI)*, pages 627–636, Tehran, Iran, March 2019. IEEE.

59. Jian Su, Wenrui Huang, Xiaoyun Qu, Tao Zhang, Lei Wang, Shuangqing Xu, Xing Yan, and Junhua Ma. Analysis of energy efficiency evaluation indexes for energy internet. In *Proceedings of IEEE Sustainable Power and Energy Conference (iSPEC)*, pages 2860–2864, Beijing, China, November 2019. IEEE.

60. Zhou Su, Yuntao Wang, Qichao Xu, Minrui Fei, Yu-Chu Tian, and Ning Zhang. A secure charging scheme for electric vehicles with smart communities in energy blockchain. *IEEE Internet of Things Journal*, 6(3):4601–4613, June 2019.

61. Panggih Sudarmono, Deendarlianto, and Adhika Widyaparaga. Energy efficiency effect on the public street lighting by using led light replacement and kwh meter installation at dki Jakarta province, Indonesia. *Journal of Physics: Conference Series*, 1022, May 2018.

62. Binod Vaidya and Hussein T. Mouftah. IoT applications and services for connected and autonomous electric vehicles. *Arabian Journal for Science and Engineering*, 45:2559–2569, April 2020.

63. Ovidiu Vermesan, Lars-Cyril Blystad, Roberto Zafalon, Alessandro Moscatrlli, Kai Kriegel, Randolf Mock, Reiner John, Marco Otella, and Pietro Perlo. Internet of energy: Connecting energy anywhere anytime. In Meyer G. and Valldorf J., editors, *Advanced Microsystems for Automotive Applications*, pages 33–48, Berlin, Heidelberg, 2011. Springer.

64. Kun Wang, Jun Yu, Yan Yu, Yirou Qian, Deze Zeng, Song Guo, Yong Xiang, and Jinsong Wu. A survey on energy internet: Architecture, approach, and emerging technologies. *IEEE Systems Journal*, 12(3):2403–2416, September 2018.

65. Yixiao Wang, Lei Hou, Kun Su, Ankang Miao, Bobo Chen, and Zhengwen Zhang. Research on integrated energy management and control system architecture based on local energy internet. In *Proceedings of IEEE 3rd Conference on Energy Internet and Energy System Integration (EI2)*, pages 6–11, Changsha, China, November 2019. IEEE.

66. Yewen Wei, Shuailong Dai, Jialin Yu, Shuo Wu, and Jiayu Wang. Research on status and prospects of battery energy storage stations on energy internet. In *Proceedings of IEEE 3rd Information Technology, Networking, Electronic and Automation Control Conference (ITNEC)*, pages 964–969, Chengdu, China, March 2019. IEEE.

67. Yiqun Xie, Jayant Gupta, Yan Li, and Shashi Shekhar. Transforming smart cities with spatial computing. In *Proceedings of IEEE International Smart Cities Conference (ISC2)*, pages 1–9, Kansas City, MO, USA, September 2018. IEEE.

68. Yu-Sheng Yang, Shih-Hsiung Lee, Guan-Sheng Chen, Chu-Sing Yang, Yueh-Min Huang, and Ting-Wei Hou. An implementation of high efficient smart street light management system for smart city. *IEEE Access*, 8:38568–38585, February 2020.

69. Yun Yang, Zhuangzhi Fang, Fanpeng Zeng, and Peizhe Liu. Research and prospect of virtual microgrids based on energy internet. In *Proceedings of IEEE Conference on Energy Internet and Energy System Integration (EI2)*, pages 1–5, Beijing, China, November 2017. IEEE.

70. Ping Yi, Ting Zhu, Bo Jiang, Ruofan Jin, and Bing Wang. Deploying energy routers in an energy internet based on electric vehicles. *IEEE Transactions on Vehicular Technology*, 65(6):4174–4725, June 2016.

71. Haihang Zhou, Jianguo Yao, Xue Liu, and Haibing Guan. Enabling energy storage in internet datacenters: Requirements, standards, and opportunities. *IEEE Internet Computing*, 20(5):66–70, October 2016.

72. Jizhong Zhu, Pingping Xie, Peizheng Xuan, Jin Zou, and Pengfei Yu. Renewable energy consumption technology under energy internet environment. In *Proceedings of IEEE Conference on Energy Internet and Energy System Integration (EI2)*, pages 1–5, Beijing, China, November 2017. IEEE.

73. Yayun Zhu, Jiye Wang, and Kehe Wu. Open system interconnection for energy: A reference model of energy internet. In *Proceedings of IEEE International Conference on Energy Internet (ICEI)*, pages 314–319, Beijing, China, May 2017. IEEE.

74. Yongqiang Zhu, Na Zhao, Shaoqian Zhang, and Wanjun Wang. Research on modes of energy utilization in regional energy internet. In *Proceedings of IEEE International Conference on Energy Internet (ICEI)*, pages 38–42, Beijing, China, May 2018. IEEE.

# 5 IoE Applications for Smart Cities

*Manju Lata*
Chaudhary Bansi Lal University, Haryana, India

*Vikas Kumar*
Chaudhary Bansi Lal University, Haryana, India

## CONTENTS

## 5.1 INTRODUCTION

Modern developments in renewable energy systems, power control systems, telecommunication tools, storage techniques and tools, information systems, cyber security and Internet of Things (IoT) have led to the initiation of advance concept of Internet of Energy (IoE) [1]. This is referring to an integrated infrastructure of dynamic network built upon the inter-operable and standard communication protocols that interrelate the energy network through the Internet enabling energy units. In general, IoE signifies the standard shift of grid systems from centralized power generation and one-dimensional power flow into more reliable, secure, flexible, sustainable, and efficient energy networks [2]. The purpose of the IoE is to make sure that a robust system is designed for extensive intelligent monitoring and control of energy through the internet. All planners, developers, designers, and engineers are seeking applications of smart technologies to attain the objective of lowest feasible energy cost with zero environmental impact, with 100% renewable scheduled to build life cycle [3]. A number of comprehensive applications for IoE based smart technologies have been presented in the Figure 5.1. Smart energy technologies are next generation technologies having a great future. IoE based building energy management system supports the reduction of $CO_2$ and energy consumption by exchanging the information of energy supply and demand in support of sustainable intelligent buildings [4]. The applications of IoE can transform the conventional cities into smart cites and conventional buildings into smart buildings, which are ecologically and financially feasible throughout the use of smart energy systems and technologies [5]. To attain sustainable energy future smart energy technologies include the renewable sources, PE converter, smart plug-and-play interfacing, Smart home appliance, smart metering technology, smart lighting, distributed generation, energy storage and net metering, smart building energy management system, smart energy router, smart grid and smart heating, ventilation and air conditioning (HVAC) technologies etc. Large scale implementation of the technologies can prevent the power blackouts in near future.

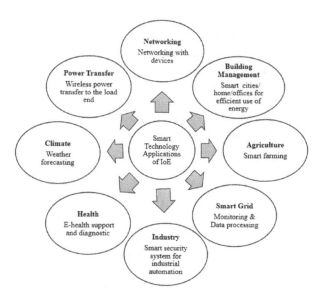

**Figure 5.1**    Smart technology applications of IoE

In addition sensors are using to bring together data that derived to particular software to be evaluated enabling real-time monitoring of the utensils, which has brought a range of profits with 5% decrease in downtime and a 25% diminution in operations costs and maintenance [6]. The World Energy Council described that there will be a lot of change in the 21st century in digital services based on the renewable energy. This renewable energy area will be worth by \$89bn in 2030 [7]. According to the International Energy Association (2011), the Global electricity requirement is estimated to increase with more than two-thirds by 2035 [8]. Furthermore, it is expected that about 40% of the overall Global energy consumption will be by the buildings and this will represent one-third of total greenhouse gas (GHG) release [9]. Wells et al. [10] have also presented an analysis on zero energy building (ZEB) with the requirement of different mobile wireless communication technologies or cellular network technologies. The ZEB concept is expected to have equal consumption and production of energy, which can significantly reduce the energy cost and demand.

## 5.2 ENERGY CHALLENGES IN IOE

With the increasing use of social economy the cities and countries have brought forth many concerns regarding the traffic congestion, severe environmental based greenhouse gas, consumption of energy and resources, extreme expansion of the population, and other issues [11]. Even though cities cover-up about 3% of the earth's surface, they consume 75% of the energy resources and produce nearby 80% of greenhouse gas [12]. These issues are acutely obstructing the rapid developments of modern society and metropolitan area. There is the great requirement to control the energy consumption intelligently and efficiently by the use of effective energy consumption reduction policies and smart technologies. Therefore, the smart cities must be planned by analyzing the various scenarios and strategically taking-up the energy issues. This become the greatest significance with amalgamate of social resources, endorsing industrialized development, and increasing speed of commercial and profitable transformation [13]. Smart city uses a number of information & communication technologies and these technologies, when efficiently used can lead to smart city life with low carbon economy and reducing the ecological impact, while improving the delivery of services. Additionally, these smart applications and technologies with the integrations of tools and techniques offer cost-effective and inventive solutions to the growing various challenges faced by modern society and metropolitan area [13]. However, regardless of the numerous benefits of smart city development plans, number of energy based challenges remains when it move towards the deployment because of exclusive and sustainable city requirements and contrary interpretations of deployment notions. There are a number of energy based challenges faced by the smart cities, in the present scenario:

### 5.2.1 RELIABILITY AND SCALABILITY

A reliable Internet of Energy (IoE) based energy management system can focus for the end user's attention, accordingly, can make sure the accomplishment of the sustainable development objective. Incorrect information, design and decision in IoE based energy management system has become a major issue in the modern society and smart cities that can cause increased disorder and given up for lost and damage [3]. Additionally, the energy generation level diverges depending on the weather stipulation, and offered power network system endures from unpredictability and unreliability due to the lack of fault analysis, efficient monitoring, and maintenance system [14]. Though, the power supply to the end user of smart city wants to be secure and reliable with satisfactory power supply. An efficient energy storage system with renewable sources can supply continuous power to the consignment with balancing power by grid and maintenance the power quality unaffected [15]. Consequently, selection of substance into the energy storage system, intellectual power infrastructure, monitoring and control system for the IoE structural design can speed up the growth of an efficient energy management system. Scalability signifies the system stability when latest applications, services, or devices are attached to the accessible system [16]. The power quality of the system should be ensured to handle all worst conditions. Also, the users would like to attain latest services consistently; hence the IoE based energy management system of smart cities should be very much scalable to support the future enhancement.

## 5.2.2   SECURITY AND PRIVACY INTENDED FOR DATA ACCESS

Data collection and analysis is a basic component of the energy management system of the smart city. In this point, different types of data, for example perpetuation or alternative plans of the utensils, human resource modifications, billing information, meter data, and spending of the energy consumption are collected frequently. This data is subsequently used to recognize the energy efficient buildings in energy management system like: cooling, heating, and lighting. Currently, a lot of data is generated from the sub-meters, smart meters, and sensors deployed to evaluate the performance of smart cities building. Therefore, selection of tools, classification of circuit, equipment assignment, verification of data and metering equipment should be reliable enough to support the efficient metering system. There is no common security standard specifically designed for IoE (in cooperation of wired and wireless network) based energy management system [17]. At this time, there is the requirement of advanced security to make sure that this system bring together a lot of information and utilizes several keys for user name and passwords [18]. In accordance with verification and validation, privacy, integrity, reliable and unaffected data recording and tracking, save and replace without missing data are the requirements of security in support of the application [17, 19]. Furthermore, the use of Internet through smart grid can enhance the incidence of a cyber-attacks and this can become a major threat to nationalized security [20]. Consequently, developments on privacy and security [21], in support of IoE based energy management system of smart city can be the excellent increment in economy to extend the quality of life of smart cities.

## 5.2.3   COST AND EXPENDITURE

The costs of IoE based energy management system of smart city consist of the costs of energy, operations, smart appliances, materials of energy storage system, building or reconstruction costs, along with the technological and maintenance costs. Earlier research of the amalgamation of renewable sources signified that the integration costs of solar energy diverge from medium to high [22], though the wind integration [23], and geothermal energy costs [24], are still high. Therefore, the technologies cost has a tough cause on IoE based energy management system. Additionally, the low cost of materials of energy storage system designed for storing the energy of renewable sources to make sure the accessibility can be a good option in support of the system [25]. Hence, an appropriate stability between the costs of IoE based energy management system and greenhouse gas emission should be maintained to make certain an efficient energy management to make the sustainable smart city [25]. Consequently, accomplishing the sustainable and cost-effective IoE based energy management system with the amalgamation of renewable sources is a challenging issue in support of further advancement of this technology [26]. Community instruction and advance research on ecological factors can transform the tendency in this regard [27].

### 5.2.4 CLIMATE CONDITIONS

Weather and geological location are the most susceptible challenge into the IoE based energy management system of smart cities [28]. Cooling and heating, Internet supply, telecommunications, ventilation, GHG emission, and thermal comfort-ability are considerably persuaded via the climate change [28]. When temperature differs, then energy consumption based on the load also differs. Thus, climate is becoming a significant determinant in support of planning an efficient energy management system for smart city [29]. Also, the wireless devices possibly will not work proficiently in insensitive atmosphere. Wind velocity and solar radiation also differ quickly that can be vital in support of power supply of the smart city systems. Consequently, energy efficient and climate sensitive energy management system of smart city is still a major challenge for upcoming augmentation.

### 5.2.5 LEGISLATION

Legislators or representatives face some significant issues including verdict workable strategies regulating stakeholders, to set free financial growth and sustaining reimbursement in support of the city's population and authorizing development in research and advance assets [30]. Receiving the personal data of the participants and matching corresponding trade-offs, that also become the greatest issue for the legislators or policymakers. General public is always bothered concerning the privacy of their personal data and movements. The mechanism of a variety of cameras and sensors on smart street lights in IoE based energy management system gives them a feel of continually being watched via the government [3]. Furthermore, the data composed by the development followers is the central part of smart projects that is able to lead a dynamic pricing model, balancing the investment cost and offering profits is sustaining the production process. Besides the privacy challenges, securing funding to start a development project and making sure that the resources sustain through the project, stay behind the most complicated challenges for smart cities [13]. Public-private corporations become the most admirable investment category used to control the financial challenges. Legislators in every county of a globe are responsive for funding and interoperability challenges in IoE based energy management faced by smart cities, subsequently formulating regular attention with the project partners remains a big challenges.

### 5.2.6    EDUCATION AND ENGAGEMENT OF CITIZENS

Education is an additional approach to mitigate the anxiety amongst the citizens in smart cities. Technology based education programs willingly accessible in IoE, and smart citizens have engaged aggressively taking benefit of latest technologies, along with governments of city can communicate the inherent reimbursement of these projects [30]. Consequently, there are only some cities selected for the smart cities' assignments of education to make the same level of development. Some already have the essential infrastructure, service, and assets model [31]. Education based smart city model is quite new that poses numerous challenges. Whereas, some cities have initiated few smart projects to get better life of city, for instance better air quality, citizen empowerment, and smart power grids [13]. Such city is working in IoE based energy management system together with entrepreneurs, residents, large companies and startups to make a new city of next day [3].

### 5.2.7    INFRASTRUCTURE AND CAPACITY BUILDING

Smart Cities of IoE based energy management system make use of sensor technology for improving the quality of life of citizens. With the increasing level of sensors, expertise for a robust connectivity also becomes the requirement of success. But, it is inadequate via the city's budget. Consistent with RootMetrics, exposure and consistency across the entire city becomes the key to introduction any programmes of successful smart city [3, 30]. Complex and valuable infrastructure is concerned in maintaining and installing the sensors for successful smart city. Although there is a biggest challenge related to power, will it be powered by involving solar energy, hard-wiring, or battery operation or a combination of all three? Some cities have already faced challenges by shifting the conventional infrastructure, for instance the transportation tunnels, steam pipes, underground wiring, and installing speedy internet [30]. Broadband wireless service is growing, even if there are still regions in foremost cities where right to use is restricted. Installing innovative sensors and additional developments cause temporary though still provoking difficulties on behalf of citizens existing in these cities. Moreover, Building the citizen capacity on behalf of greatest amount of smart cities is not a simple effort and most determined projects are overdue because of lack in quality of manpower at every level. Through the use of modern infrastructure related technologies, the most important objective is to work together with citizens, academic, businesses, and other stakeholders to solve the largest amount of challenges [13, 17].

## 5.3  RESOLUTIONS TO IOE ENERGY CHALLENGES

Reducing the energy consumption, maximizing energy efficiency by minimizing losses, and minimizing its effect on the climate are challenging issues in energy management systems of smart cities [30]. This is essential that these challenges should be overcome by coming up with a standard sustainable city. The Global smart city market size was valued at USD 567.45 billion in 2013 and is expected to reach around $1422.57 billion in 2020 [32]. The sustainability effects of a smart city become complex to segregate and determine, due to inherent procedures to the smart city itself. When a smart city development project put together the solutions, procedures, performers, and contradicts incompetence's connected with silo-oriented society, it becomes almost unfeasible to determine energy sustainability as a distinct objective [13]. Within the smart city projects, sustainability work together with other objectives, where all of them are influencing each other and producing probable analysis and irregular effect of energy challenge. However, as the city population increases and metropolitan slump disseminates, problems with financial and social development are often magnified. These challenges not only have an effect on a city's quality of life, but also add strain on conventional infrastructure, increase the need for energy efficiency severity with resource protection. Smart city technologies can offer to the governments of city through a huge infrastructure buffer that facilitates them undergo and overcome these challenges in future [17, 30]. Due to increasing share and growth of smart technologies of the world, the energy consumption is also increasing. This increment process of energy consummation leads to negative impacts on a global scale and the smart cities themselves also suffering from this impact, and demanding the significant energy saving materials with cost effectiveness [33]. As each city has its significant way of life, infrastructure, and funding procedures, technology implementation can differ within various ways. Conversely, in this way it is not constantly possible to depend on other established smart city projects to do something as a blueprint for accomplishment. Although, the metropolitan areas are continually work together with the external globe where the regions situated. There is no global or local efficiency, exclusive of collaboration with a variety of regions, together with dealing on energy transactions between low production and high consumption areas. This is estimated that around 58% globe population will live in metropolitan areas by 2025, putting up a big challenge to the infrastructure of cities [33, 34]. With the world-wide focus on the development of smart cities, these challenges are not very easy to be taken up in the smart infrastructure as well. Smart city poses substantial challenges, together with climate change, overpopulation, ecological quality and right to use energy. Smart city development should reconsider how to sustainably provide the essential services at a reasonable price to the residents. Foremost city consume around 65% of existing key energy and add up on behalf of approximately 70% of greenhouse gas emissions, generally to make available energy in support of cooling, heating, transport, and lighting. In accordance with the World Health Organization (2016) report, a big challenge waits for metropolitan areas, where large metropolitan populations require clean air to take breaths [34]. Thus, an important challenge for the smart cities can be taking up the climate change and deterioration of air quality. Environmental and energy challenges become the key point, to extent the limitation of taking to all aspects of smart city development. IoE including the digitalization such as sensors, big data, and IoT will facilitate the efficient and optimal management of the whole system of the city [35]. With IoE based energy management system, smart cities can have some economic and practical resolutions:

*Ecological impact:* Reduction in $CO_2$ footprint becomes the foremost driver after the augmentation of sustainable and smart cities. Increasing storage and energy efficiency, improving traffic conditions and better waste management can offer specific benefits.

*Optimized water and energy management:* Smart water management and grids are returning themes of smart cities. Monitoring of potable water and consumption of energy makes sure the quality of tap water and accessibility of energy across the city.

*Transportation:* Efficient and clean transportation of services, goods, and citizens is crucial. Within the anticipation of mobility optimization, several cities are turning toward the smart technologies to mitigate traffic congestion and present users by way of concurrent updates.

*Protection:* Security is a main concern on behalf of all cities. The increased speed in expansion of smart cities must facilitate the urban monitoring of their citizens via CCTV cameras with the identification. Additionally, High-tech CCTV cameras are prepared with movement and smoke detectors with fire alarms.

## 5.4   SMART APPLICATIONS OF IOE

The concept IoE based smart city is becoming smarter in relation to the development of digital technology and communication [36]. The considerable enhancement for digital technology is drive to massive business prospective in support of the IoE, since the entire technologies and devices are able to integrate and communicate with each other via Internet. The state of affairs is always towards the ever-increasing quality of service provided to the citizens, whereas making an inexpensive benefit on behalf of the city management towards the operational costs [37]. The most important component, where IoE can play a considerable role includes:

*Smart Energy:* Smart energy within IoE supplies an inventive method in favor of power distribution, grid communication and monitoring, energy storage. IoE can facilitate the automatic allocation of energy according to the requirement that how many units are to be relocated, when and where. Energy consumption observation executed on every level, as of local individual appliances up to state and worldwide level [38].

*Smart Health:* IoE has a good quality in favor of various medical applications and a variety of wearable devices accessible in market such as Pulse rate, Heart rate, BAN devices, smart wristwatch, and etc. Those are extensively used to make a diagnosis and monitoring the patient's constantly. With the large number of ubiquitous wearable devices in healthcare systems obtain benefits of varied mobile networks (Wi-Fi, cellular network, Bluetooth, ZigBee, WiMAX, and etc.) [20]. The processing and computations are made by cloud on the basis of IoE with powerful computational servers collecting the data sensed via devices after that analyze and process the information [39].

*Smart Home:* Radio access networks (Wi-Fi, Bluetooth, WiMAX etc.) become the basic building blocks of home automation. Nowadays, most of devices are the electronic devices that become use in homes with communication ability. All the smart things in smart-home work automatically open are closed based on the sensed information via the sensors for example lights, windows, etc. A number of associations are functioning on this, providing smart home through the use of technology allow users using single device to manage all electronic appliances. IoE make it feasible to have the automated home applications [20].

*Smart Transport and Mobility:* The automobiles with embedded technology of IoE are able to communicate using Internet with a number of innovative applications, which carry latest functionalities to support the safe and convenient transport. Within the perspective of the concept of smart transport and mobility interrelated with Internet of Energy (IoE), there is a significant upcoming trends in support of smart applications and technology [20].

Therefore, the Smart city with energy management system of IoE involve the sustainable city development by building foremost enhancements in terms of city direction, and elucidations to a large number of problems such as traffic management, energy utilization, parking, pollution control, power plants, waste disposal and water supply networks [40]. The IoE can significantly contribute to an inventive breed of smart city that develops the sustainable information and communication technologies covering the education, health, quality of life, reduce costs, along with the operation of metropolitan services in support of citizens [3, 30]. IoE facilitated the modern technologies on behalf of energy region that should be re-evaluated to rally the varying necessities of the future [41]. All mechanism of the upcoming energy system such as consumers, producers, storage, grids, and etc., can be coupled with the standardized open building. Smart city stands out in favor of its predominance with smart lifestyle, management, housing, mobility, and smart economy. The foremost objective is to bring together hi-tech modernization with social, economic, and environmental issues of the city of tomorrow. There are a number of examples of cities leading the way in urban sustainability using IoE with the application of IoT [42, 43].

*Singapore:* Smart nation program has been launched by Singapore addressing mobility issues that making significant investments in phased traffic lights, road sensors, and smart parking regarding to managing energy and green innovation. This city hosts the clean-tech park including green buildings and green industries.

*Barcelona:* This city became an avant-garde city that has been produced the Urban Lab. This is a smart city enterprise, where corporations offer sustainable thoughts to develop the life of local residents. With the use of sensors, this city is managing lighting, traffic lights, and green spaces. Additionally mobility, smart waste collection, and traffic become the main concern.

*San Francisco:* This connected city plan to facilitate the citizens accessing data designed to build the easier life of citizens and report them in the field of biodiversity, mobility, and health. That extremely concerned in sustainable improvement together with waste recycling, the city is dedicated to 100% renewable electricity in favor of urban services.

*Oslo:* This is able to focus on smart lighting in the Norwegian capital; where nearby 10,000 streetlights have been prepared by sensors to correct the brightness consistent with lighting and season's requirements. The main objective is reducing the electricity consumption with 70%. Furthermore, this city has desires to limit the car use in urban regions, and determined the mobility strategy derived from public transport has been commenced.

*London:* This city has been concerned for smart and sustainable development. Regarding to mobility, this city became a predecessor with initiating a toll for cars and CCTV for observation services. In this city pedestrians are available all through the drive with interactive terminals. With the use of IoE techniques various aspect of smart city has been developed rapidly, and all connected devices and modern networking technologies have become increasingly important and serving a better quality of life with ultimate security [44]. In this distributed system, there is the need to transform old grids into an Internet of Energy. Therefore, the lots of useful reasons about smart city signified the realization of the IoE [45]. Following are some role that defined the how IoE is evolved to solve the problems of smart city.

*Delivering safe and secure energy within smart city* Smart IoE technology is serving to solve problems giving consumers greater insight and control over how much energy they consume. Various consumers around the globe already have been well-known with smart meters concept. The devices and procedure are designed to communicate straightforwardly with home's gas or electricity meter, and energy provider. This immediate connection signifies that consumers are able to observe accurately resulting cost on energy consumption [46]. So, consumers are able to obtain an informed resolution on what to use and how they can perform more cost-efficiently. Simultaneously, utilities present exact billing, consistent with immediate energy consumption, and carefully manage and balance supply and demand. The renewed highlight on energy efficiency and possibly overwhelming the effects of climate change that is also led to growing use of renewable energy sources (wind, solar, or wave power) [47]. By the way of using and integrating renewable energy sources into consumer homes, it can be the case that lots of costumer rely on own energy production, with possible excess of energy to be store or sell back into the grid [46]. Obviously, energy grids become a significant part of state level infrastructure and keep them safe is vital. Thus, involving smart grids have need of robust security to secure against cyber-attacks or hacking. Although the smart energy ecosystem is still developing and the entire stakeholders push onward and work together to make sure its future strength, stability and protection.

*Smart City Corporations and Ecosystems* To see all the solutions at scale, there is the need the type of collaboration between the authorities' public and private businesses has to develop. In ancient times, the conventional correlation was: buyer equal to public authorities along with one-time supplier equal to private business [45, 48]. This type of correlation cannot propose the sustainable solutions and corporations at the moment. To come up the city problems must be followed the improvement of well-built surroundings of partners, means various dissimilar suppliers with corresponding skills and services that can supply resources to innovate mutually with city [49]. This type of environment has the requirements to embrace stakeholders from dissimilar surroundings, for instance local universities, local authorities, public or private organizations, setup and some key companies. All stakeholders can include the incremental value to a particular part of the value chain. Individual business cannot provide everything, and cities can't faith for each solution to one provider. Corporations and co-innovation based on the IoE management system can be pilot, plan, and scale a smart city solution [48]. IoE can revolutionizes the correlations and build ecosystems that is the obvious requirement in favor of new business models.

*Smart city plans for reducing air pollution* With the power of IoE including IoT and AI, smart cities hold the capacity to identify the granular level and immediate major air pollution problems, the reason is being that who is inflated and what it intended for the citizens [45]. By means of all these concurrent insights the city supervisors are able to take revealing decisions concerning how to undertake such complexity and how to prioritize the outlays. There are the numbers of air quality sensors, which are able to place in public means of transportation, smart equipment such as smart benches, smart lights or something else that is able to linked, for instance bus stations, bikes, rubbish bins, and etc. A number of ecological information is widely accessible, and a few is able to offer to the limited councils in short or without any cost on replace to something else as license using city space or include sensors on city furniture [49]. Actually, when data from sensors is computing the quality of air that shared by unspecified mobile data from the mobile operator's network. Cities are able to preparation to build new perambulator streets, electric vehicle chargers, new biking routes, or parking spaces depend on the quality of air. Therefore, both the collected data from mobile phones and sensors is significant [45]. In another way, several tools are being developed to optimize the decisions based on concurrent data or automate processes derived from particular events. Additionally at present, the cities are able to modify the actions on region level and obtain dissimilar procedures for each region, in place of substitute in the same way in favor of big areas of the city.

*Centralized grids and smarter grids supply smart city* Since the energy system develops more difficult, the old centralized grid is harassed to sustain. This issue deteriorates as more decentralized units, for instance solar panels are amalgamated into the system [48]. Such as, the German power grid thought about a role model in favor of developed countries, almost by 100 percent accessibility [49]. Though, the energy transition in the direction of renewable already has overstated it significantly. Recently there is the enormous development in the number of times network operatives included intervene to become stable grid. Patrons finish up paying in favor of these interventions by the use of charges on their electric bills. It is the matter of constrained requirement within developing countries. The increase of phones, computers and other devices in developed countries is also fueling demand intended for electricity. Evidently, for the solution in terms of the power from renewable energy sources can be offered with a climate-neutral alternative. Although, the fundamental shift is appeared with incredible technical challenges, that involve a quite new mode of managing energy system [50]. Managing energy system, there is the requirement of power grids smarter. With the use of digital technology, the upcoming smart grids will attach individual consumers and producers to form an Internet of Energy (IoE) that make certain a consistent, stable, and secure power supply.

*Dealing with mountains of data* Volumes of data are increasingly getting complicated as distributed energy site is producing large amount of data. To solve this complexity, a large number of networked devices come in picture. Billions of devices are now attached to internet, if we consider the global scenario. Therefore, more than a few attached devices has grown rapidly and is extensively predictable to keep on at a swift pace. Gartner estimated that numerous attached devices have jumped to above 20.4 billion, from 8.4 billion in 2017. Ericsson has a dissimilar mode of counting, excluding anticipated development at a related scale. The corporation observed that the attached devices growing to above 25 billion by 2020 furthermore to 29 billion by 2022, as compared to 17 billion in 2017 [51]. The universal data area of expertise is expected to rise to 175 zettabytes by 2025, concerning 10 times the 16.1 zettabytes produced in 2016, consistent with marketplace investigate firm IDC (1 zettabyte is equivalent to 1 trillion gigabytes) [51, 52]. This foremost challenges that how to use this data to develop systems is solving by modern technologies like IoE in collaboration with IoT, cloudlet, big data analytics, and others.

*An Internet of Energy for all* According to the history of Internet, each new applicant within the network enhance the potentials and may contribute to support the whole system [38]. Same analogy can be used in the favor of Internet of Energy. Its importance increase while an effortless power grid develops into an all-inclusive energy network. Within such a network, excess electricity from one source being capable of supply the foundation in support of producing other forms of energy to other sectors. As electricity is transformed into heat or used to extract heat, it may supply the foundation in favor of providing mobility, or when it is utilized to create methane or hydrogen, totally flexible resources of energy become accessible. All networking technologies increase latest significance in the procedure of change. More willingly than simply distributing and transmitting electricity, the power grid increase into an innovative globe shifting network that is the Internet of Energy.

## 5.5  CONCLUSION

The requirement of smart services in increasing day by day because of the numerous projects towards improving the infrastructure and smart tools and techniques. Cities are facing a number of energy related challenges, including the data security and privacy, infrastructure, cost and legislation. Each one of these must be taken-up strategically to offer a solution to the growing needs of the ever-increasing population. Consumers, researchers, developers, home and industrial sectors are always looking for new and optimal solutions to increase the energy efficiency and reduce the harmful effects of consumption. Therefore, IoE supports the organizations and countries to control the increasing demand of energy, enabling energy infrastructure to manage the peak and low consumption demand. Along with this, IOE can automate this process and take-up many other challenges existing with the urban infrastructure and can certainly offer a pathway to the emergence of energy efficient smart cities.

## REFERENCES

1.  Hersent, O., Boswarthick, D., & Elloumi, O. 2012. "The internet of things: Key applications and protocols", John Wiley & Sons, www.wiley.com
2.  Shahinzadeh, H., Moradi, J., Gharehpetian, G. B., Nafisi, H., & Abedi, M. 2019. "Internet of Energy (IoE) in smart power systems." In 2019 5th Conference on Knowledge Based Engineering and Innovation (KBEI): pp. 627-636.
3.  Hannan, M. A., Faisal, M., Ker, et al. 2018. "A review of internet of energy based building energy management systems: Issues and recommendations." IEEE Access, 6, pp. 38997-39014.
4.  Wei, C., & Li, Y. 2011. "Design of energy consumption monitoring and energy-saving management system of intelligent building based on the Internet of Things." In 2011 International Conference on Electronics, Communications and Control (ICECC): pp. 3650-3652.
5.  Jindal, A., Kumar, N., & Singh, M. 2020. "Internet of energy-based demand response management scheme for smart homes and PHEVs using SVM." Future Generation Computer Systems, 108, pp. 1058-1068.
6.  Bassirian, P., Moody, J., & Bowers, S. M. 2017. "Analysis of quadratic Dickson based envelope detectors for IoE sensor node applications." In 2017 IEEE MTT-S International Microwave Symposium (IMS): pp. 215-218.
7.  Oliver F. 2019. https://www.nesgt.com/blog/2019/05/what-is-the-internet-of-energy (accessed on May 26, 2019)
8.  International Energy Association. 2011. "World Energy Outlook 2011 Factsheet: How will global energy markets evolve to 2035." IEA, Paris, France, 2013, pp.1-5.

9. Huang, Z., Lu, Y., Wei, M., & Liu, J. 2017. "Performance analysis of optimal designed hybrid energy systems for grid-connected nearly/net zero energy buildings." Energy, 141, pp. 1795-1809.

10. Wells, L., Rismanchi, B., & Aye, L. 2018. "A review of Net Zero Energy Buildings with reflections on the Australian context." Energy and Buildings, 158, pp. 616-628.

11. Puiu, D., Barnaghi, P., Toenjes, R., et al. 2016. "City-pulse: Large scale data analytics framework for smart cities." IEEE Access, 4, pp. 1086-1108.

12. Mohanty, S. P., Choppali, U., & Kougianos, E. 2016. "Everything you wanted to know about smart cities: The internet of things is the backbone." IEEE Consumer Electronics Magazine, 5(3), pp. 60-70.

13. Kuru, K., & Ansell, D. 2020. "TCitySmartF: A comprehensive systematic framework for transforming cities into smart cities." IEEE Access, 8, pp. 18615-18644.

14. Jaradat, M., Jarrah, M., Bousselham, A., Jararweh, Y., & Al-Ayyoub, M. 2015. "The internet of energy: smart sensor networks and big data management for smart grid." 56, pp.592-597.

15. Cao, J., & Yang, M. 2013. "Energy internet–towards smart grid 2.0." In 2013 4th International Conference on Networking and Distributed Computing: pp. 105-110.

16. Khajenasiri, I., Estebsari, A., Verhelst, M., & Gielen, G. 2017. "A review on Internet of Things solutions for intelligent energy control in buildings for smart city applications." Energy Procedia, 111, pp. 770-779.

17. Wang, K., Yu, J., Yu, Y., et al. 2017. "A survey on energy internet: Architecture, approach, and emerging technologies." IEEE Systems Journal, 12(3), pp. 2403-2416.

18. Al-Fuqaha, A., Guizani, M., Mohammadi, M., Aledhari, M., & Ayyash, M. 2015. "Internet of Things: A survey on enabling technologies, protocols, and applications." IEEE Communications Surveys & Tutorials, 17(4), pp. 2347-2376.

19. Xu, T., Wendt, J. B., & Potkonjak, M. 2014. "Security of IoT systems: Design challenges and opportunities." In 2014 IEEE/ACM International Conference on Computer-Aided Design (ICCAD): pp. 417-423.

20. Yang, Y., Wu, L., Yin, G., Li, L., & Zhao, H. 2017. "A survey on security and privacy issues in Internet-of-Things." IEEE Internet of Things Journal, 4(5), pp.1250–1258.

21. Jia, G., Han, G., Jiang, J., Sun, N., & Wang, K. 2015. "Dynamic resource partitioning for heterogeneous multi-core-based cloud computing in smart cities." IEEE Access, 4, pp. 108-118.

22. Cheng, T. C., Cheng, C. H., Huang, Z. Z., & Liao, G. C. 2011. "Development of an energy-saving module via combination of solar cells and thermoelectric coolers for green building applications." Energy, 36(1), pp. 133-140.

23. Al Busaidi, A. S., Kazem, H. A., Al-Badi, A. H., & Khan, M. F. 2016. "A review of optimum sizing of hybrid PV – Wind renewable energy systems in Oman." Renewable and Sustainable Energy Reviews, 53, pp. 185-193.

24. Tian, X., Yang, M. J., Zhao, J. W., He, S. M., & Zhao, J. 2015. "A study on operational strategy of ground–source heat pump system based on variation of building load." Energy Procedia, 75, pp. 1508-1513.

25. Loukaidou, K., Michopoulos, A., & Zachariadis, T. 2017. "Nearly-zero energy buildings: Cost-optimal analysis of building envelope characteristics." Procedia Environmental Sciences, 38, pp. 20-27.

26. Liang, H., Abdrabou, A., & Zhuang, W. 2014. "Stochastic information management for voltage regulation in smart distribution systems." In IEEE INFOCOM 2014-IEEE Conference on Computer Communications: pp. 2652-2660.

27. Deuble, M. P., & de Dear, R. J. 2012. "Green occupants for green buildings: The missing link?" Building and Environment, 56, pp. 21-27.

28. Ciscar, J. C., Feyen, L., Soria, A., et al. 2014. "Climate impacts in Europe – The JRC PESETA II project." Vol. 26586, pp. 1-157.

29. Santamouris, M. 2016. "Innovating to zero the building sector in Europe: Minimising the energy consumption, eradication of the energy poverty and mitigating the local climate change." Solar Energy, 128, pp. 61-94.

30. Angiello, G., Carpentieri, G., Pinto, V., Russo, L., & Zucaro, F. 2014. "Review pages: Planning for smart cities dealing with new urban challenges." TeMA. Journal of Land Use, Mobility and Environment, 7(3), pp. 333-358.

31. Coe, A., Paquet, G., & Roy, J. 2001. "E-governance and smart communities: a social learning challenge." Social Science Computer Review, 19(1), pp. 80-93.

32. San Francisco. 2016. https://www.globenewswire.com/news-release/2016/04/07/826691/0/en/Smart-Cities-Market-Expected-To-Reach-USD-1-422-57-Billion-by-2020-Grand-View-Research-Inc.html (accessed on April 07 2016)

33. https://www.ifpenergiesnouvelles.com/article/smart-city-energy-challenges-facing-sustainable-cities (accessed on Oct. 2018)

34. World Health Organization. 2016. "Ambient air pollution: A global assessment of exposure and burden of disease", [Available at]: https://apps.who.int/iris/handle/10665/250141

35. Hossein Motlagh, N., Mohammadrezaei, M., Hunt, J., & Zakeri, B. 2020. "Internet of Things (IoT) and the energy sector." Energies, 13(2), 494.

36. Jindal, A. 2018. Data Analytics of Smart Grid Environment for Efficient Management of Demand Response (Doctoral dissertation, Thapar Institute of Engineering and Technology Patiala).

37. Arasteh, H., Hosseinnezhad, V., Loia, V., et al. 2016. "Iot-based smart cities: a survey." In 2016 IEEE 16th International Conference on Environment and Electrical Engineering (EEEIC): pp. 1-6.

38. Vermesan, O., Blystad, L. C., Zafalon, R., et al. 2011. "Internet of Energy connecting Energy anywhere anytime", In Advanced microsystems for automotive applications 2011, pp. 33-48, Springer, Berlin, Heidelberg.

39. http://www.nist.gov/el/upload/CPS (accessed on June 10 2017)

40. Mahapatra, C., Moharana, A. K., & Leung, V. 2017. "Energy management in smart cities based on internet of things: Peak demand reduction and energy savings." Sensors, 17(12), p. 2812.

41. Kafle, Y. R., Mahmud, K., Morsalin, S., & Town, G. E. 2016. "Towards an Internet of Energy", In 2016 IEEE International Conference on Power System Technology (POWERCON): pp. 1-6.

42. Saleem, A., Khan, A., Malik, S. U. R., Pervaiz, H., Malik, H., Alam, M., & Jindal, A. 2020. "FESDA: Fog-enabled secure data aggregation in smart grid IoT network." IEEE Internet of Things Journal, 7(7), pp. 6132-6142.

43. Bibri, S. E. 2018. "The IoT for smart sustainable cities of the future: An analytical framework for sensor-based big data applications for environmental sustainability." Sustainable Cities and Society, 38, pp. 230-253.

44. Ejaz, W., Naeem, M., Shahid, A., Anpalagan, A., & Jo, M. 2017. "Efficient energy management for the internet of things in smart cities," IEEE Communications Magazine, 55(1), pp. 84-91.

45. Neirotti, P., De Marco, A., Cagliano, A. C., Mangano, G., & Scorrano, F. 2014. "Current trends in Smart City initiatives: Some stylised facts." Cities, 38, pp. 25-36.

46. Kunold, I., Kuller, M., Bauer, J., & Karaoglan, N. 2011. "A system concept of an energy information system in flats using wireless technologies and smart metering devices." In 6th IEEE International Conference on Intelligent Data Acquisition and Advanced Computing Systems (2): pp. 812-816.

47. Shrouf, F., Ordieres, J., & Miragliotta, G. 2014. "Smart factories in Industry 4.0: A review of the concept and of energy management approached in production based on the Internet of Things paradigm." In 2014 IEEE International Conference on Industrial Engineering and Engineering Management: pp. 697-701.

48. Schaffers, H., Komninos, N., Pallot, M., Trousse, B., Nilsson, M., & Oliveira, A., 2011. "Smart cities and the future internet: Towards cooperation frameworks for open innovation", In The future internet assembly, pp. 431-446, Springer, Berlin, Heidelberg.

49. Cedrik N. 2018. https://www.greenbiz.com/article/5-problems-internet-energy-can-solve (accessed on February 27 2018)

50. Jindal, A., Kronawitter, J., Kuhn, R., Bor, M., de Meer, H., Gouglidis, A., Hutchison, D., Marnerides, A.K., Scott, A., & Mauthe, A. 2020. "A flexible ICT architecture to support ancillary services in future electricity distribution networks: an accounting use case for DSOs." Energy Informatics, 3(1), pp. 1-10.

51. Reinsel, D., Gantz, J., & Rydning, J. 2018. "The digitization of the world from edge to core", IDC White Paper. [Available at]: https://www.seagate.com/files/www-content/our-story/trends/files/idc-seagate-dataage-whitepaper.pdf

52. Patrizio, A. 2018. "IDC: Expect 175 zettabytes of data worldwide by 2025", Network World. [Available at]: https://www.networkworld.com/article/3325397/idc-expect-175-zettabytes-of-data-worldwide-by-2025.html

# 6 IoE Design Principles and Architecture

*Rania Salih Abdalla*
Dept. of Electronics Engineering,
Sudan University of Science and Technology, Sudan

*Sara A. Mahbub*
Dept. of Electronics Engineering,
Sudan University of Science and Technology, Sudan

*Rania A. Mokhtar*
Dept. Computer Engineering,
Taif University, Al-Taif, KSA; Dept. of Electronics Engineering,
Sudan University of Science and Technology, Sudan

*Elmustafa Sayed Ali*
Dept. of Electronics Engineering,
Sudan University of Science and Technology,
Sudan; Dept. of Electrical and Electronics Engineering,
Red Sea University, Sudan

*Rashid A. Saeed*
Dept. Computer Engineering, Taif University,
Al-Taif, KSA; Dept. of Electronics Engineering,
Sudan University of Science and Technology, Sudan

## CONTENTS

## 6.1   INTRODUCTION

The term Internet of Energy (IoE) was firstly inspired in the third industrial revolution by (Jeremy Rifkin, 2008) [1], which refers to an electricity solution for power flow and bidirectional information in an internet-style, which is also known as energy internet and it considered as a smart grid extension [2, 3]. IoE has recently become an interesting field and attracts a lot of researchers and developers from almost all developing countries as in Europe, US, Germany, China, and Japan [4, 5]. IoE acts as a cloud network where power sources with embedded and distributed intelligence are interfaced to smart grid and mass of consumption devices like smart buildings, appliances, and electric vehicles.

In fact, smart metering technology has a vital role in IoE [8]. In addition to its application in smart automation applications (see Figure 6.1), it enables the ability of consumption and energy sensing with either electronic or electromechanical components, that helps consumers in management and optimization of energy resources in both commercial and residential various applications [5]. This rapid increase of IoE application in modern life applications and industry led to innovation of IoE standards which will be discussed later in this chapter, in addition to IoE architectures, interoperability, privacy, and security.

The need for continuous development in energy management has different factors indeed. Starting by development in energy generation techniques from traditional ways to modern renewable techniques. Then distribution and transformation development will be also required to compatible forms with that developed energy generation techniques. Also, the adoption of technology in various real live applications including civilian and manufacturing environments causes a massive increase in number of users and devices with numerous processes and services that consumes energy rabidly [6, 7]. This development introduces more challenges in centralized infrastructures of energy production and distribution and rises the need to develop modern and sophisticated solutions to meet both energy efficient and applications demands requirements.

The negative environmental impacts of traditional energy systems such as fossil fuel energy generation turns the focus of energy producers to renewable energy as a promising way for green life applications and the best choice for business demands [8]. Moreover, the development of renewable energy techniques such as solar and wind energy are main factors to develop an integrated energy platform that has the decentralization ability and compatibility with current techniques in use.

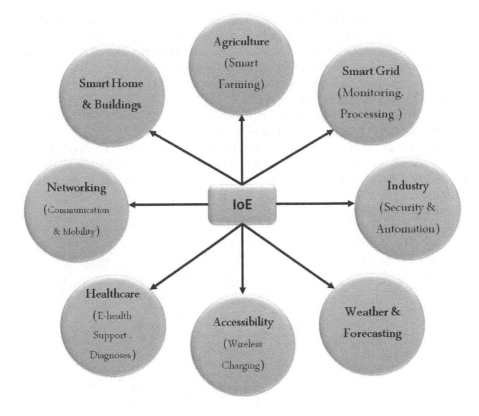

**Figure 6.1** IoE applications

IoE concept rises to meet these requirement of modern life applications, and has been applied and developed within various domains rapidly. Recently, it got an attention of researcher and developers to develop an integrated framework that integrates IoE with security perspectives and other modern technologies such as cloud computing, fog computing, and blockchain [9]. This integration of IoE application in modern life applications and industry technologies led to innovation of IoE standards which will be discussed later in this chapter, in addition to IoE architectures, interoperability, privacy, and security.

## 6.2 IOE ARCHITECTURE MODELS

Traditionally, energy systems deploy generation, transmission, and distribution. Then IoE is invented as an ICT solution to add a communication layer or functionality that integrates all system components together in end-to-end fashion, while providing other system services. This integration of IoE platform includes various sectors from system management to data security and development tools. By integration of intelligent end devices, networking, real-time capability, and integrated applications for business and mobile devices portal [10, 44]. Technically, in order to achieve such integration required by IoE platform proposal, the most practical suitable tactic is the well know service-oriented architectures (SOA), where networking communication protocols are used to provide service delivery between different system components. Figure 6.2, shows the SOA architecture-based IoE.

SOA avoids point-to-point interfaces between single applications, by implementing one single interface for each application to the integrated platform, then it can be used to dynamically address usage to other application. Actually, this unique feature is pointed with a developed approach known as extended SOA introduced by (M. G. Keen, 2006) that must introduce more flexibility and scalability to the whole system by integrating the individual components to an intelligent end device as referred to by intelligent grid devices (IGD) [35].

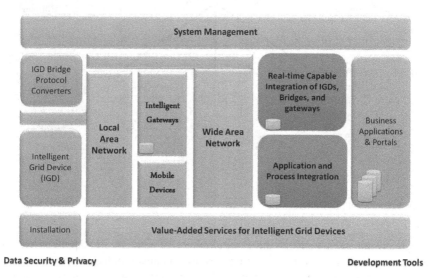

**Figure 6.2**  IoE deployment sectors for SOA architecture [6]

## 6.2.1 AN EMS-BASED ARCHITECTURE

An energy management system (EMS) refers to the power grid control center, which takes charge of monitoring and mission management, beside of its important role in power system operation's safety and stability, it represent a core factor in development of IoE architecture [11]. The EMS-based architecture in (E. Hossain, 2012), developed an EMS system based on home sensor and actuator network (HSANET) installed at a customer location [12]. At service provider location data and service center (DSC) is installed.

Major structure components employed by EMS are shown in Figure 6.3, where basic information infrastructure is represented by the HSANET which include a

ZigBee-based home sensor network and ZigBee-internet home gateway and control center. The home sensor network (HSN) is acting as a user interface, which is responsible of energy flow control, sensing, and monitoring. Home gateway and control center (HGCC) is acting as an interfacing device between the service provider and the customer to provide intelligence energy management operations.

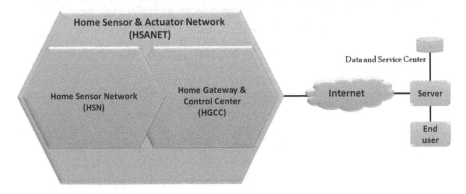

**Figure 6.3**  Basic IoE EMS-based architecture

Improvement in EMS is emerging and reaches all energy grid operation levels from generation to distribution to enable optimal use of traditional or renewable energy sources as in decentralized generation, while maintaining system stability and service quality [13, 45]. The architecture above represent a basic form of EMS-based architecture of IoE that was more suitable with centralized energy management, while there were various developed relative architectures are introduced specially with growing of decentralized energy generation management as shown in Figure 6.4.

The advancement in EMS-based IoE architecture followed different paths in development, where an architecture in Figure 6.4 that presented by (Y. R. Kafle et al. 2016) involves an additional functionalities and features such as in different energy generation sources, energy storage, enhanced data center, and smart metering technology [2]. This type of architectures considers the decentralized energy generation and management that may include common renewable energy forms like solar and wind, in such architectures other forms are also usable such as hydroelectric, geothermal, biomass etc.

Figure 6.4 above shows an advanced EMS based architecture shows more features and functions are enabled for both energy sources and load, including renewable energy sources, distributed energy storage, plug-in electric vehicles, domestic and industrial "prosumers", etc. Prosumers concept introduced in this architecture refers to mixture of producer and consumer or as called an active consumer who would participate in energy production by their own [12, 13]. Such architecture mainly depends on internet for all operations required for control and monitoring.

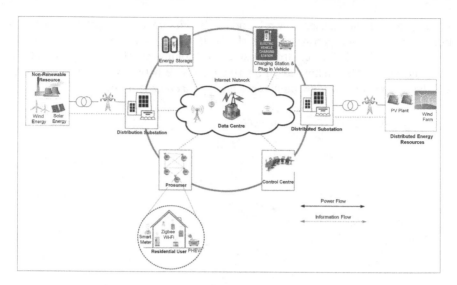

**Figure 6.4**  An advanced EMS-based IoE architecture [2]

Currently, by 2020 more development on EMS-based architecture are introduced by multiple institutes, the most common one that combines major of these developments is referred to by Micro-Service based EMS architecture [14], the authors here addresses major factors on development of EMS as also expressed main facilities and issues with an existing EMS as also they compared their architecture against the service-oriented architecture energy management systems (SOA-EMS) based architecture in different perspectives.

An illustration of the proposed MS-EMS based architecture is shown in Figure 6.5. The figure shows the major development part here is achieved in software platforms and devices integration, in addition to the usage of virtual private cloud (VPC) that discussed in multiple developing architectures [15, 16], the software development environment used in this architecture depends on C# and MATLAB programming languages in addition to representational state transfer (REST) application programming interface (API) [17, 18]. Concurrently with these developments based on cloud computing technology, an advanced architecture is developed with present of transactive energy management systems and referred by Fog-based IoE architecture.

## 6.2.2  A FOG-BASED ARCHITECTURE

The challenge of addressing the optimal use to achieve both operational and business objectives by utilization of companies and customers in distributed energy resources is still an ongoing issue, which initiated by the transactive energy (TE) methodology. The grid wise architecture council (GWAC) [19] defined TE as the methodology of an electric power system management by using of economic or market-based

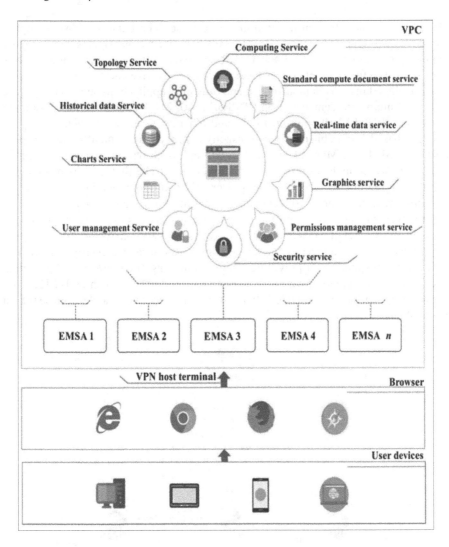

**Figure 6.5** Micro service EMS-based IoE architecture [12]

constructs in generation, consumption or electric power flow, to provide market and control functions jointly. The system connectivity map that includes flow of electrical energy through physical components or points which referred to by transactive nodes (TN).

TNs are controlled in real time based on economic impulses or incentive, insurance of control system scalability, which is achieved by the decentralized transactions and information exchange between the TNs. A transactive node uses transactive incentive signal (TIS) which represent foretoken transferred electric energy cost, and the transactive feedback signal (TFS) to represent forecasted total power flow at a the particular transactive node. The balance between supply and demand is achieved by exchanging of that transactive signals between the neighboring transactive nodes. Each TN echoes to the system situations among decisions related to the conductance of local assets. Transactive energy has been excessively deployed in many demonstration and industrial applications such as the pacific northwest smart grid demonstration (PNSGD) [20, 21], Pacific Northwest National Laboratory (PNNL) Voltron project [22], and other applications as given in.

A fog-based IoE architecture for transactive energy management systems is proposed by (M. H. Y. Moghaddam, 2018), the approach design model is shown in Figure 6.6, which includes three different layers [23]. The first layer is responsible of providing an interface between the power grid and customers through the home gateways, by transaction of collected energy consumption data of the customer. The second layer is responsible of providing low latency services to the end users, by locating the Fog nodes at the network edge to act as an energy market server agent that represented by the retail energy market server of the transactive energy system. The third layer is responsible of providing a high computing environment and perpetual data storage, by supporting various communication protocols such as HTTP, Constrained Application Protocol (CoAP), and Open automated demand response (Open ADR) alliance.

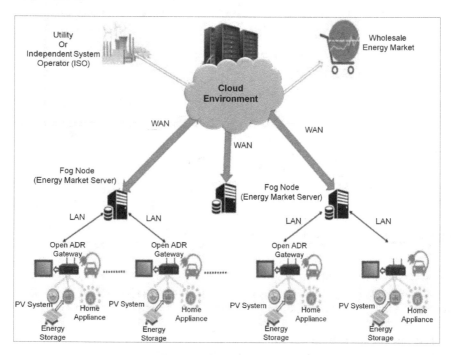

**Figure 6.6**    A fog-based IoE architecture [29]

The demonstration of the proposed Fog-based IoE architecture is implemented with simple and available components as shown in Figure 6.7(a), starting with the IoT node includes:

Electricity meter or current sensor.
Electric control switching relay.
Temperature, humidity, gas, light and pressure sensors.
Wi-Fi Module.

The Home Gateway is demonstrated by a raspberry-Pi3 microprocessor-based computer, and the system computing environment is deployed with the "Things Speak" fog and cloud computing environment that connects the home gateway through an active internet connection. Different communication protocols and standards involved in the proposed architecture includes 802.11, IP, TCP, UDP, COAP, HTTP, and Open ADR as illustrated in Figure 6.7(b) which deployed within the actual hardware system components.

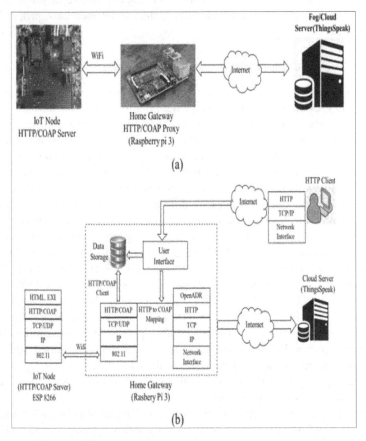

**Figure 6.7** (a) Fog-based IoE demonstration (b) communication protocols [29]

## 6.3   EMBEDDING INTELLIGENCE IN IOE DESIGN

As development of emerging technology-based systems are rapidly growing, and adoption of its applications in various real-world fields, data analysis and storage platforms are also requiring a suitable and sophisticated framework to integrate that development and offer reliability and scalability of systems specially in real-time applications and such as required by IoE methodology. Computational environment and tools are found to meet that challenge by enabling more computation and storage capabilities that can affect system performance concurrently.

Development in web-based applications represents the main factor behind computation tools development, but also finds its way in other applications such as networking, communication, healthcare, industry, energy, etc. As mentioned in previous sections IoE concept occurrence and architecture evolution is mainly depends on computational environment growth as in cloud computing, fog or edge computing, and latest emerging technology known as blockchain technology [30]. All of these technologies use a common principle which is the decentralization and distribution of service nodes somehow, but may differs in its locations, properties, and functionalities.

### 6.3.1   CLOUD COMPUTING

Cloud computing represents an evolutionary version of computational tools that consists an improved and reliable management platform for data warehousing, data analysis, and monitoring. Cloud computing providers promises to maintain data availability and reliability of all data related operations by providing an evolutionary infrastructure that includes multiple ways for data warehousing and data aggregation framework, in addition to modern developed functionalities that required for an emerging applications such as data monitoring and visualization, big data analysis, real-time and low latency processing, and diversity of computation resources [30]. Cloud computing introduced various solutions for services provider by offering four main services are; infrastructure as a service (IaaS), platform as a service (Paas), software as a service (SaaS), and backend as a service (BaaS).

Cloud computing upgraded traditional computation environment by enabling an on-demand and availability of data by distribution of services into multiple locations from central servers, here virtual machine (VM) had played the main role of enabling multiple and different computation capabilities on a single hardware, also software development plays an important role in building application programming interfaces (API) for data interaction and management. Additionally, cloud computing offered an excellent integration and compatibility with IoT applications and platforms by enabling multiplicity for various IoT platforms and higher system bandwidth capabilities as required by IoT applications.

A typical Energy Internet architecture featuring integrated information and energy infrastructures highlighting the connectivity between energy routers and cloud-based data centers introduced by Huawei is shown in Figure 6.8. The application of cloud computing in IoE introduces more opportunity and solutions in today challenges between demand and suppliers. (R. Fu et al. 2017) discusses two major techniques that would help in development of energy internet or IoE which are the big data and clouding computing technologies, with a focus on the platform architecture that is particularly adaptive to handle a variety of characteristics of energy industry data [24].

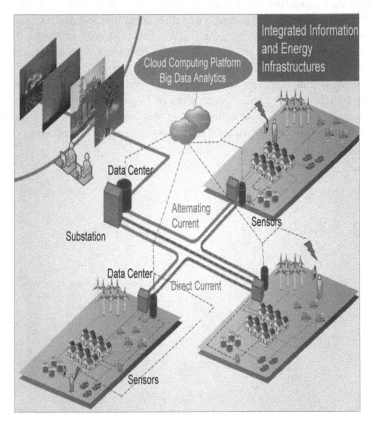

**Figure 6.8**  Typical cloud-based IoE architecture [32]

In study presented by (F. Ma et al., 2017), authors discussed the concepts of cloud computing, cyber security scheme, and system architecture design scheme of the developed cloud-computing platform in power system planning studies [14]. The study proved that the adopting cloud computing enables to effectively meet many computing needs in power system and provides cost-effective benefits without compromising cyber security and data privacy that are equally concerning to such power organizations.

An implementation of cloud computing technique for power system data acquisition, storage, and analytics are discussed by (M. Feng et al., 2015) at New England [15]. From two use cases presented in this work, they state that adoption of cloud computing can is useful for system operators i.e. ISOs by enabling an efficient and flexible supplement to traditional IT infrastructure and encouragement of the coop-eration across different system operators with the assurance of privacy and security.

## 6.3.2  FOG COMPUTING

Generally, Fog Computing concept grows with spreading of IoT applications, it dif-fers from cloud computing by adding a new concept called an edge computing which are majorly devices at an edge between supplier and customer gateways. These edge devices represent the new layer added to bridge terminals with the system compu-tation core as shown in Figure 6.9. Simply, the Fog computing never eliminates the cloud computing but represent an evolutionary version that is complement cloud computing by adding edge node. As growing of Fog computing and its applications in developing systems, IoE also is involved in this development and various propos-als and demonstrations are introduced as discussed in previous sections.

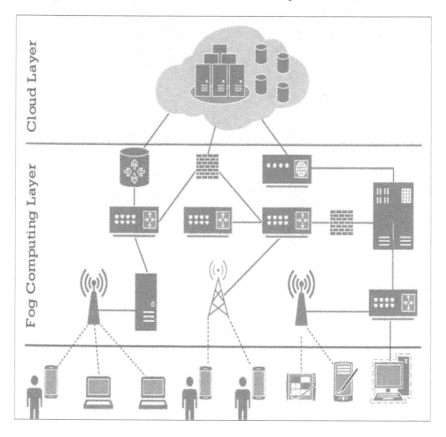

**Figure 6.9**  Fog computation architecture [33]

Current research efforts, describe applications where fog computing is benefi-cial and identified future challenges that remain open to bring fog computing to a breakthrough [30]. (K. Shahryari et al., 2017) introduced bridges fog domain and cloud in a smart grid gateway for scheduling devices/appliances by creating a priority queue. This mechanism enables to perform demand side management dynamically [26]. The queue is affected by the consumer importance and policies, in addition to the status of energy resources. Later (M. H. Y. Moghaddam et al., 2018) present a fog-based IoE architecture for transactive energy (TE) management systems that is fully described as shown in Figure 6.6 and 6.7 in previous sections [23].

### 6.3.3 BLOCKCHAIN

Blockchain in general represent an innovated data transaction framework with more security and reliability by introducing additional security and data integrity procedures. In larger, blockchain technology introduces the features of automation, immutability, public ledger facility, irreversibility, decentralization, consensus and security [27, 47]. The use of emerging new technologies such as blockchain is encouraged according to the security and privacy requirement in IoE centralized structure by providing an autonomous, decentralized, and cooperative principle using smart contract and cryptography as defined by its revolutionary architecture (see Figure 6.10). Blockchain integration with IoE framework can effectively introduce more security and transparency, in addition to time and energy efficiency with low operational costs. Recent extensive surveys on application of blockchain in different IoT and smart grid demonstrations can be found in [27, 29], that discusses challenges and solutions of blockchain technology functions and its adoption with energy application.

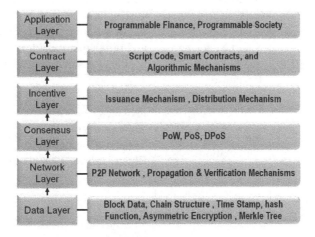

**Figure 6.10**   Peer-to-Peer (P2P) architecture of blockchain

## 6.4   IOE STANDARDS

Technical instructions and specification of most ICT applications are standardized by Institute of Electronics and Electrical Engineering (IEEE), International Electrotechnical Commission (IEC), etc. [46]. To grant demands and reliability of renewable energy, IEEE has established multiple standards and guideline to cover ICT aspects across many geographic and standards development organization (SDO) boundaries for smart energy systems which are summarized in Table 6.1 which illustrates the main IEEE standards for smart energy systems.

**Table 6.1**
**Main IEEE Standard for Smart Energy Systems**

| Standard Series | Name | Description |
|---|---|---|
| IEEE 2030 | Smart Grid Interoperability | Defines data flows in electric power. Reliable and secure bidirectional flow. Identifies the necessary communication infrastructure for electric vehicles. |
| IEEE 1901 | Networking and Communications | Addresses broadband/narrowband over power line. Addresses many other aspects of networking. |
| IEEE 1686, IEEE C37.240, and IEEE 1711 series | Cyber Security for Smart Grid | Addressing cyber security for intelligent electronic devices and substation automation. |
| IEEE 170X, IEEE 1377, IEEE 2030.5, and IEEE 1901 series | Smart Metering and Demand Response | Addresses smart grids communication and protocols. Provides smart energy profiles for smart metering functionality. |
| IEEE 1815 (DNP3), IEC/IEEE 61850-9-3, IEEE C37.238, IEEE C37.118 series | Substation Automation | Provides: Time protocol. Synchronization work. Electric power system communication. |
| IEEE 2030.1.1 | Electric Vehicle Charging | Specify: The design interface of electric vehicles. Direct current and bidirectional chargers that utilize battery electric vehicles as power storage devices. |

### 6.4.1  IEEE 2030 STANDARD

IEEE 2030 standard is intended to but the roadmap to achieve smart grid interoperability, by establishing the smart grid interoperability reference model (SGIRM) that models the framework of engineering principles to apply smart grid interoperability upon all electric power system components. These visions are looking beyond the year 2030 and are predicting how the smart grid will evolve for each of these technology-focused areas [37]. The package of IEEE 2030 standard series includes many recommendations and guides for design, implementation, and evaluation sectors of smart grid systems as summarized in Table 6.2.

### 6.4.2  IEEE 802.15.4G

Design consideration of smart grid requires modern solutions for networking and communication operations, Wi-Fi and Zigbee wireless communication was the available suitable standards as also demonstrated with some basic smart grid applications, but more specifications and procedures are required, IEEE 802.15.4g standard is developed to meet that requirements. IEEE 802.15.4g describes the standard wireless communication platform for smart utility service, architecture of IEEE 802.15.4g communication platform as shown in Figure 6.11 [35]. Both orthogonal frequency division multiplexing (OFDM) and multi-regional frequency shift keying (MR-FSK) operating in sub-1GHz frequency are used in the physical layer. The selection of those PHYs depends on the types of applications, and both can be controlled by the MAC layer. The MAC layer can control both PHYs, we do not operate them simultaneously. Implementation of MAC hardware accelerator and the OFDM and MR-FSK modem is performed with an emerging FPGA technology [33, 35]. The FSK and OFDM RF blocks are both implemented in ASICs which are fabricated through CMOS 0.18um process [38]. The MAC and application software are running on a 32-bit microcontroller.

### 6.4.3  IEEE 21450 AND IEEE 21451

International organization for standardization, and the international electrotechnical commission (ISO/IEC/IEEE 21450) standard provides a common foundation for members of the (ISO/IEC/IEEE 21451) series of international standards to be able to exchange and make use of information. It realizes the functionalities that are to be performed by a transducer interface module (TIM) and the joint characteristics for all TIM based devices. It also defines the formats for transducer electronic data sheets (TEDS) and the set of commands to simplify the configuration and control of the TIM as well as reading and writing the data used by the system. To facilitate communications with the TIM and with applications, application programming interfaces (APIs) are defined [38].

**Table 6.2**

**IEEE 2030 Standard Recommendations and Practice Guides**

| Standard | Description |
|---|---|
| IEEE 2030 | Guide for: Smart grid interoperability of energy technology and information Technology operation with the electric power system (EPS) End-use applications, and loads |
| IEEE 2030.1.1 | Standard technical specifications of a DC quick charger and bi-directional charger for use with electric vehicles |
| IEEE 2030.2 | Guide for the interoperability of energy storage systems integrated with the electric power infrastructure |
| IEEE 2030.2.1 | Guide for: Design, operation, and maintenance of battery energy storage systems Both stationary and mobile applications integrated with electric power systems |
| IEEE 2030.3 | Standard test procedures for electric energy storage equipment and systems for electric power systems applications |
| IEEE P2030.4 | Draft guide for control and automation installations applied to the electric power infrastructure |
| IEEE 2030.6 | Guide for the benefit evaluation of electric power grid customer demand response |
| IEEE SA - 2030.5™ | Ecosystem steering committee |
| IEEE 2030.5 | Standard for smart energy profile application protocol standard |
| IEEE 2030.7 | Standard for the specification of microgrid controllers |
| IEEE 2030.8 | Standard for the testing of microgrid controllers |
| IEEE 2030.9 | Recommended practice for the planning and design of the microgrid |
| IEEE 2030.100 | Practice for implementing an IEC 61850 based substation communications, protection, monitoring, and control system |
| IEEE 2030.101 | Guide for designing a time synchronization system for power substations |
| IEEE P2030.100.1 | Draft monitoring and diagnostics of IEC 61850 generic object-oriented status event (GOOSE) and sampled values-based systems |
| IEEE P2030.102.1 | Draft Standard for Interoperability of Internet Protocol Security (IPsec) Utilized within Utility Control Systems |

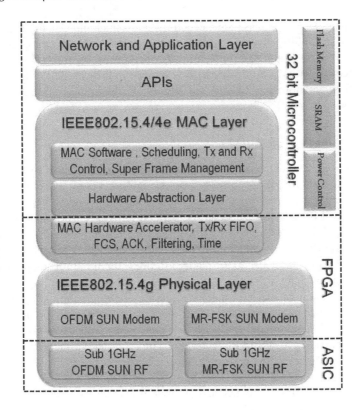

**Figure 6.11**    IEEE 802.15.4G communication platform architecture

## 6.4.4   THE 4TH G -BASED LOW POWER WIDE AREA (LPWA)

Low power wide area networks are one of new networking solutions that developed to provide an alternative for traditional short-range wireless networks such as Wi-Fi. This is makes it a good communication technique for the future of IoE that by introducing low implementation costs and energy efficiency. Modern solutions of IoE in both residential and industrial areas require clean and efficient energy solutions which introduced by LPWA networks that enables wider coverage and longer battery life.

Technically, there is two types of LPWA networks; licensed and unlicensed LPWA, unlicensed LPWA using unlicensed spectrum and it specified for customized applications. Licensed LPWA is more commercial and uses licensed spectrum as in Narrow Band Long-Term Evolution (NB-LTE) and LTE-M LPWA networks which are introduced within 4thG or LTE the evolutionary cellular networks [30]. LPWAN is an emerging network technology for IoE and M2M applications which suitable for long-rang performance and cellular M2M networks.

The integration of IoT and M2M devices connected by the LPWAN technologies will generate a global IoT in anywhere and anytime to interact with any kind of environment. Issues such as range, penetrability of thick objects, long battery life, security and scalability all will be adequate by the LPWAN technologies. To support the future growth and development of the Internet of Energy (IoE), the mobile industry together with the third Generation Partnership Project (3GPP) has standardized new class technologies: LTE-M and NB-IoT for low power wide area network applications.

## 6.5  IOE INTEROPERABILITY

IEEE 2030 Standard, defined the application guide for smart grid interoperability reference model (SGIRM) to energy storage. It highlights the standard related to the information relevant to energy storage system (EES) interoperability, with energy power system (EPS). The interoperability standard provides useful industry-derived definitions for ESS characteristics, applications, and terminology that, in turn, simplify the task of defining system information and communications technology (ICT) requirements [34]. The IEEE 2030 standard defines, the three IoE major sectors of interoperability, which are the power systems interoperability, communications technology interoperability, and information technology interoperability.

For power systems interoperability, the IEEE standard known as smart grid interpretability reference model (SGIRM) presents a power systems interoperability architecture perspective (PS-IAP), which introduces a view of the EPS that "assure the production, delivery, and consumption of electrical energy". It also defines the power information logical flow between specified entities and domains through interfaces that enable intelligent control and optimization of interoperable power system components as shown in Figure 6.12.

The SGIRM also presents a communication systems interoperability architecture perspective (CS-IAP), which provides more details about the communications technologies associated with the particular aspects of energy storage. Figure 6.13 shows the CT-IAP modified to highlight pertinent ESS entities and interfaces.

For power systems IT interoperability, SGIRM provides an information technology interoperability architecture perspective (IT-IAP), which provides more definition of the methods of information exchanged between entities that interact with the ESSs. Figure 6.14 presents the IT-IAP modified to highlight pertinent energy storage systems data flows and entities.

## 6.6  IOE PRIVACY AND SECURITY

In critical infrastructures such as smart power grids and IoE, the issue of dependability and security is not fully understood yet, and privacy is an additional security objective. The IEEE Standards between 2011 and 2030 provides an overview of security issues, strategies, security requirements, risk management, security design, and countermeasures besides the standards and best practice recommendations [39]. Additionally, an end-to-end security, security by design, and security in depth are the most security concepts must be included within security conceptual model.

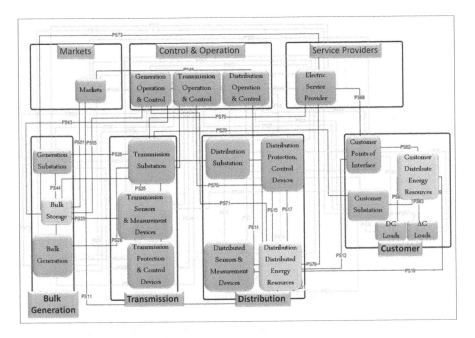

**Figure 6.12** PS-IAP modified to highlight pertinent ESS entities and interfaces

While design security in IoE is usually developed by relates to the manufacturing of individual products and the assembly of systems, solutions, and architectures. Security in depth implies the realization that any security feature by itself is breakable with enough effort, and only multiple security controls layered in a concentric way around protected assets can provide a security superior to the sum of the individual parts [40]. And an End-to-end security is safeguarding information in an information system from point of origin to point of destination. Full end-to-end security, however, would require all endpoints to support a common control security mechanism.

According to IEEE standard recommendation and standard, the IEEE 1686$^{TM}$, IEEE P37.240$^{TM}$, IEEE 1711$^{TM}$, IEEE 1402$^{TM}$, and IEC 62351 series are the best practice for cyber security introduced for IoE solutions. While the federal energy regulatory commission (FERC), recognizing the needs of energy sector, as identified in demand and response, wide area situational awareness, energy storage, electric transportation. Also, FERC identifies two cross-cutting priorities, namely, cyber security and communication and coordination, across inter-system data flows. Similarly, the department of energy (DoE) has identified that the smart grid should operate resiliently against physical and cyber-attacks and natural disasters.

## 6.6.1 IOE CYBER SECURITY

Cyber security in IoE concerns about all procedures and methods of securing communication and networking for both information data and power flow. According to

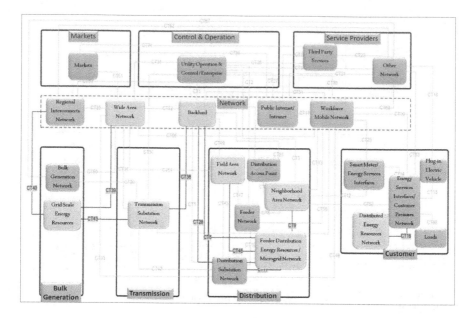

**Figure 6.13**    CT-IAP modified to highlight pertinent ESS entities and interfaces

IEEE standard, an organization has to apply analysis and risk management methods to identify the appropriate solutions to ensure the security of the distributed energy resources DERs including related systems and Smart Grid.

The security engineer has to understand the standards of information exchange, to fill the gaps that should be taken into account. That data exchange recommended standards and application programming interfaces that supportive of smart grid technologies and corresponding ISO/RTO services or products based on the information demand response (DR) models are supplied in standard publication documents by the national institute for standards and technology (NIST), The national electric sector cyber security organization resource (NESCOR), and electric power research institute (EPRI), etc.

These recommendations and standards are rich of key functional security regions, various markets, and critical infrastructure protection (CIP) cyber security and other pertinent accuracy standards. There are also other security requirements and recommendations available in documents such as north American energy standards board (NAESB), smart grid interoperability panel (SGIP) Priority Action Plan PAP19, and SGIP PAP09. However, the security gaps and other requirements issues require resolution, and improvements should be combined with the design of security for DR systems and applications.

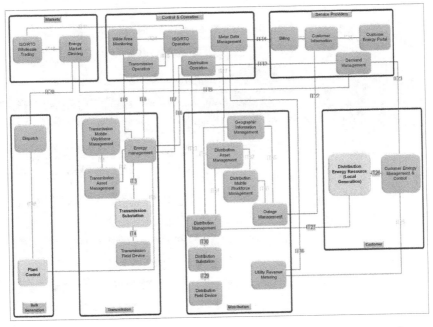

**Figure 6.14** IT-IAP modified to highlight pertinent energy storage systems data flows and entities

## 6.6.2 IOE HARDWARE SECURITY

For some specified hardware components, there is an identified standard recommendation introduces by IEEE. Such recommendations like IEEE 1686 [41] which introduces the intelligent electronic devices (IEDs) features and functions to be provided and accommodate CIP programs are also introduced in this standard. It also addresses security procedures related the access, operation, configuration, firmware revision and data retrieval from an IED. But communications for the objective of power system protection (PSP) referred to by tele-protection are not addressed in this standard. IEEE 1689 defines general requirements to protect serial supervisory control and data acquisition (SCADA) and communications to maintenance ports of remote terminal units or intelligent electronic devices.

In addition, IEEE 1689 outlines requirements to retrofit legacy control systems with security that minimizes needed changes. IEEE P1711 is a trial-use standard for a cryptographic protocol for electric sector substation communications using serial links [42]. This standard applies to SCADA systems and, in particular, the communications between the SCADA master and the intelligent electronic devices (IED). The IEEE 1402 guide identifies and discusses security issues related to human intervention during the construction, operation (except for natural disasters), and

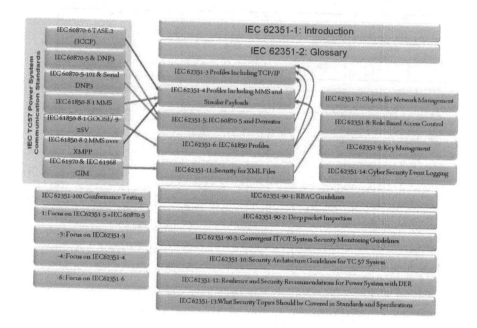

**Figure 6.15** The IEC 62351 security standards and their interrelationships with the IEC TC57 standards

maintenance of electric power supply substations [43]. It also documents methods and designs to mitigate intrusions. While the IEC 62351 family of standards as reviewed in Figure 6.15, depicts the architecture of a secure power system and standardizes its protocols and components.

## 6.7 CONCLUSION

The emerging adoption of IoT technology in energy systems and applications is referred to by IoE, which involves an integration of information communication technology with different energy or power systems components from generation, transformation, storage, distribution, and end-devices consumption. Figure 6.16 shows the IoE design layers and architecture at the different levels of energy network. Development in ICT is rabidly growth and features are introducing consequently, cloud computing, fog or edge computing, and blockchain are modern solutions for internet-based applications. Those ICT technologies can introduce an efficient, reliable, secure, and flexible infrastructure for IoE real-world implementations, which defines the IoE architecture. Standards, recommendations, and best practices are the important documents required for optimal and efficient design and deployments of modern technology applications in various perspectives such as interoperability, scalability, security and privacy. That standards are introduced by an international associations and institutes.

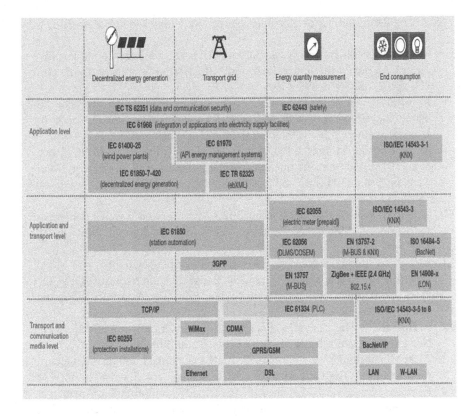

**Figure 6.16**  Energy network levels IoE design architecture

## REFERENCES

1. J. R. 2008. The Third Industrial Revolution. Engineering & Technology, vol. 3, pp. 26-27.
2. Y. R. Kafle, K. Mahmud, S. Morsalin, and G. E. Town. 2016. Towards an internet of energy. IEEE International Conference on Power System Technology (POWERCON), pp. 1-6.
3. J. Cao and M. Yang. 2013. Energy internet – towards Smart Grid 2.0. 2013 4th International Conference on Networking and Distributed Computing (ICNDC), pp. 105-110.
4. A. Q. Huang, et al. 2011. The future renewable electric energy delivery and management (FREEDM) system: The energy internet. Proceedings of the IEEE 99.1, pp. 133-148.
5. C. Block. 2016. Internet of Energy, ICT for Energy Markets of the Future. Available at: http://www.iese.fraunhofer.de/content/dam/iese/en/media center/documents/BDI_initiative_IoE_us-IdE-Broschuere_tcm27-45653.pdf

6. Z. Liu. 2015. Global energy interconnection. China Electric Power Press, Beijing.

7. R. Abe, R.H. Taoka, and D. McQuilkin. 2011. Digital grid: Communicative electrical grids of the future Smart Grid. IEEE Transactions on *Smart Grid* 2.2, vol. 2, no. 2, pp. 399-410.

8. M. Moness and A. M. Moustafa. 2016. A Survey of Cyber-Physical Advances and Challenges of Wind Energy Conversion Systems: Prospects for Internet of Energy. IEEE Internet of Things Journal, vol. 3, pp. 134-145.

9. M. A. Hannan, M. Faisal, P. J. Ker, L. H. Mun, K. Parvin, T. M. I. Mahlia, et al. 2018. A Review of Internet of Energy Based Building Energy Management Systems: Issues and Recommendations. IEEE Access, vol. 6, pp. 38997-39014.

10. Z. Lyu, H. Wei, X. Bai, and C. Lian. 2020. Microservice-Based Architecture for an Energy Management System. IEEE Systems Journal, 14(4), 5061-5072, pp. 1-12.

11. J. A. Reddy and K. S. Swarup. 2003. Open Distributed Client Server Model in an Energy Management System for Power System Security. TENCON 2003. Conference on Convergent Technologies for Asia-Pacific Region, vol. 3, pp. 981-985

12. E. Hossain, Z. Han, and H. V. Poor. 2012. Smart Grid Communications and Networking. Cambridge University Press, Cambridge.

13. A. A. Munshi and Y. A. I. Mohamed. 2018. Data Lake Lambda Architecture for Smart Grids Big Data Analytics. IEEE Access, vol. 6, pp. 40463-40471.

14. F. Ma, X. Luo, and E. Litvinov. 2016. Cloud Computing for Power System Simulations at ISO New England—Experiences and Challenges. IEEE Transactions on Smart Grid, vol. 7, pp. 2596-2603.

15. M. Feng, L. Xiaochuan, Q. F. Q. Zhang, and E. Litvinov. 2015. Cloud computing: an innovative IT paradigm to facilitate power system operations. IEEE Power & Energy Society General Meeting, pp. 1-5.

16. D. Yang, H. Wei, Y. Zhu, P. Li, and J. Tan. 2019. Virtual Private Cloud Based Power-Dispatching Automation System—Architecture and Application. IEEE Transactions on Industrial Informatics, vol. 15, pp. 1756-1766.

17. D.J. Hammerstrom_ 2013. Pacific Northwest smart grid demonstration transactive coordination signals. PNWD-4402 Rev X," Battelle—Pacific Northwest Division, Richland, WA, USA.

18. J. Haack, B. Akyol, B. Carpenter, C. Tews, and L. Foglesong. 2013. Volttron TM: An agent platform for the smart grid. International Conference on Connected Vehicles and Expo (ICCVE), pp. 1367-1368.

19. J. Hagerman_ 2015. EERE & Buildings to Grid Integration. DOE Building Technologies Office, Washington.

20. F. Wurtz and B. Delinchant. 2017. "Smart Buildings" Integrated in "Smart Grids": A Key Challenge for the Energy Transition by Using Physical Models and Optimization with a "Human-in-the-Loop" approach. Comptes Rendus Physique, vol. 18, pp. 428-444.

21. S. Widergren, J. Fuller, C. Marinovici, and A. Somani. 2014. Residential transactive control demonstration. IEEE Innovative Smart Grid Technologies (ISGT) Conference, pp. 1-5.
22. D. Forfia, M. Knight, and R. Melton. 2016. The View from the Top of the Mountain: Building a Community of Practice with the GridWise Transactive Energy Framework. IEEE Power and Energy Magazine vol. 14, pp. 25-33.
23. M. H. Y. Moghaddam and A. Leon-Garcia. 2018. A Fog-Based Internet of Energy Architecture for Transactive Energy Management Systems. IEEE Internet of Things Journal, vol. 5, pp. 1055-1069.
24. R. Fu, F. Gao, R. Zeng, J. Hu, Y. Luo, and L. Qu. 2017. Big data and cloud computing platform for energy Internet. China International Electrical and Energy Conference (CIEEC), pp. 681-686.
25. J. Gedeon, J. Heuschkel, L. Wang, and M. Mühlhäuser. 2018. Fog computing: Current research and future challenges. 1. GI/ITG KuVS Fachgespräche Fog Computing, At Darmstadt, Germany.
26. K. Shahryari and A. Anvari-Moghaddam. 2017. Demand side management using the internet of energy based on fog and cloud computing. IEEE International Conference on Internet of Things (iThings), pp. 931-936.
27. A. Miglani, N. Kumar, V. Chamola, and S. Zeadally. 2020. Blockchain for Internet of Energy Management: Review, Solutions, and Challenges. Computer Communications, vol. 151, pp. 395-418.
28. K. Wang, X. Hu, H. Li, P. Li, D. Zeng, and S. Guo. 2017. A Survey on Energy Internet Communications for Sustainability. IEEE Transactions on Sustainable Computing, vol. 2, pp. 231-254.
29. J. Wu and N. Tran. 2018. Application of Blockchain Technology in Sustainable Energy Systems: An Overview. Sustainability Journal, MDPI, *Sustainability*, 10(9), 3067.
30. A. Annaswamy. 2013. IEEE Vision for Smart Grid Control: 2030 and Beyond Roadmap. IEEE Vision for Smart Grid Control: 2030 and Beyond Roadmap, pp. 1-12.
31. S. Lee, B. Kim, M. Oh, Y. Jeon, and S. Choi. 2013. Implementation of IEEE 802.15.4g wireless communication platform for smart utility service. IEEE 3rd International Conference on Consumer Electronics in Berlin (ICCE-Berlin), pp. 287-289.
32. B. K. Muni and S. K. Patra. 2013. FPGA implementation of ZigBee Baseband Transceiver System for IEEE 802.15.4. In Advances in Computing, Communication, and Control. Berlin, Heidelberg, Springer, pp. 465-474.
33. S. S. Lee, B. Kim, J. Y. Kim, S. Choi, and C. Kim. 2013. An IEEE 802.15.4g SUN OFDM-based RF CMOS transceiver for Smart Grid and CEs. IEEE International Conference on Consumer Electronics (ICCE), pp. 520-521.
34. Y. Ma, A. Cherian, and D. Wobschall. 2017. A combined ISO/IEC/IEEE 21451-4 and -2 data acquisition module. IEEE Sensors Applications Symposium (SAS), pp. 1-6.

35. M. G. Keen, H. H. Chin, C. Ganapathi, D. Ghazaleh, and P. Krogdahl. 2006. Patterns: Extended Enterprise Soa and Web Services. IBM Redbooks Publication, SG24-7135-00.

36. T. Basso, S. Chakraborty, A. Hoke, and M. Coddington. 2015. IEEE 1547 Standards Advancing Grid Modernization. IEEE 42nd Photovoltaic Specialist Conference (PVSC), pp. 1-5.

37. Y. Khersonsky. 2012. Don't Re-invent the "Smart Grid"; Use IEEE Standards to Make It "Smart". IEEE PES Innovative Smart Grid Technologies (ISGT), pp. 1-6.

38. T. Wen and X. Bair. 2012. Application Research Base on System Engineering for Analyzing Smart Grid Standards. IEEE PES Innovative Smart Grid Technologies, 3(1), 78-85, pp. 1-3.

39. Elmustafa Sayed Ali, Rashid A. Saeed. 2014. A Survey of Big Data Cloud Computing Security. International Journal of Computer Science and Software Engineering.

40. S. Hameed et al. 2019. Understanding Security Requirements and Challenges in Internet of Things (IoT): A Review. Journal of Computer Networks and Communications.

41. G. N. Ericsson. 2010. Cyber Security and Power System Communication—Essential Parts of a Smart Grid Infrastructure. IEEE Transaction on Power Delivery.

42. S. Hong and M. Lee. 2010. Challenges and direction toward secure communication in the SCADA system. 8th Annual Communication Networks and Services Research Conference.

43. S. Hurd. 2007. Tutorial: Security in electric utility control systems. 34th Annual Western Protective Relay Conference.

44. A. Jindal, N. Kumar, and M. Singh. 2020. A Unified Framework for Big Data Acquisition, Storage, and Analytics for Demand Response Management in Smart Cities. Future Generation Computer Systems, vol. 108, pp. 921-934.

45. A. Jindal, B. S. Bhambhu, M. Singh, N. Kumar, and K. Naik. A Heuristic-Based Appliance Scheduling Scheme for Smart Homes. 2020. IEEE Transactions on Industrial Informatics, vol. 16, pp. 3242-3255.

46. A. Jindal, A. K. Marnerides, A. Gouglidis, A. Mauthe, D. Hutchison. 2019. Communication standards for distributed renewable energy sources integration in future electricity distribution networks. IEEE International Conference on Acoustics, Speech and Signal Processing (ICASSP), pp. 8390-8393.

47. S. Aggarwal, R. Chaudhary, G. S. Aujla, A. Jindal, A. Dua, N. Kumar. 2018. Energychain: Enabling energy trading for smart homes using blockchains in smart grid ecosystem. 1st ACM MobiHoc Workshop on Networking and Cybersecurity for Smart Cities, pp. 1-6.

# Section IV

## Machine Learning Models

# 7 Machine Learning Models for Smart Cities

*Dristi Datta*
Varendra University, Rajshahi, Bangladesh

*Nurul I. Sarkar*
Auckland University of Technology, Auckland, New Zealand

## CONTENTS

## 7.1 INTRODUCTION

From the last few decades, it is noticed that the overall population of the world keep increasing and more people are becoming interested to move in urban areas rather than rural areas because of high availability of job, good education and medical facilities and so on. To accommodate a high volume of population and maintaining the standard of citizen lifestyle are getting challenging day by day [24].

According to the reports of the UN [11], the increasing population trend is ongoing for cities, starting from 1950 when it was only 30%. However, at present, now

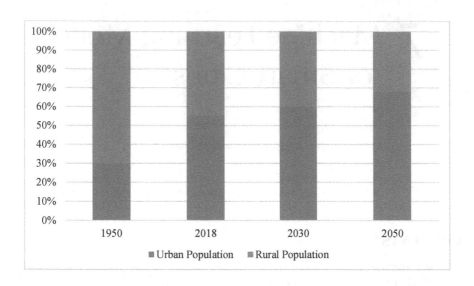

**Figure 7.1**   Growth of the urban population

it is approximately just over 55%, which is expected to reach about 68% in 2050 (Figure 7.1).

Due to the growth of the population worldwide, many cities are facing a crisis of overpopulation, shortage of resources, social and economic imbalance causing huge problems in the community. To overcome the above problems that are facing many urban areas, researchers are proposed a concept of city smart. The smart city can solve the numerous problems that are faced by conventional cities on a regular basis by efficient handling and real-time monitoring of the different parameters [23].

Smart cities (SC) are designed for the efficient and optimal delivery of facilities, as well as the optimal use of space and resources. It also focuses on improving communication between citizens at various levels. The system handles several situations in the current system and consequently, saves time and money. As technology advances rapidly, one can recognize the need for such ML models to promote and design SC. A typical smart city should have followings components [3, 5, 28]:

> *Smart Energy Management and Control:* To maximize the efficiency and profit for both stakeholders and customers, conventional power grids are needed to be converted into smart grids. The smart grid can provide low-cost electric power, real-time monitoring, maintain the continuity of power flow, and ensure better security. The addition of renewables into the smart grid makes the proper use of free sources of energy instead of high consumption of fossil fuels.

*Smart Environment Control:* This is an important element of a SC. This involves managing air pollution by continuous monitoring and sensing the level of fresh air in the environment. A large amount of sensor data can be collected and processed for an informed decision about city pollution. Waste and other environmental management can be controlled in a more effective way.

*Smart Buildings:* The high-performance buildings and urban housing can be integrated with the smart grid system for efficient use of energy to reduce energy consumption. These buildings also include a home energy management system and security solutions.

*Smart Mobility:* Fast and improved access to everywhere is possible in a smart city. This system includes traffic management, congestion management, and smart parking facilities.

*Smart Health:* This service allows patients to monitor his/her health from home linked to smart hospitals. In case of an emergency situation such as a heart attack, an ambulance will be called automatically.

*Smart Infrastructure:* These infrastructures are required for citizen's safety in the city. The system may include video surveillance cameras, internet telephony, sanitization, drainage, and various emergency support.

*Smart Governance:* The local government in the city needs to actively participate in promoting and making the city smart and its management. The city mayor and his management team are involved in city urban planning and citizen to collaborate with the project. This local body will provide key information for quality living as well as security for every citizen.

Therefore, a smart city puts various systems on the same platform to provide and control smart services to all citizens more efficiently. Figure 7.2 shows the various key components of a typical SC through a schematic diagram.

The performance of smart cities depends not only on the physical infrastructure but also on the availability and quality of data that is obtained from the various smart devices [8, 13]. The main enabler of these smart city applications is cloud computing and IoT, where everyday technology and devices are connected to network technology [21, 26]. Smart city engagement leads to a significant increase in the volume of big data across the system of various dimensions. Consequently, such a large volume of data processing and analysis become the central part of the services provided by the system. The big data phenomenon has long been characterized by the volume, speed, and data types that have been created at an increasing rate [7, 12, 20]. These big data offer the possibility of obtaining valuable information about the city from the large amount of data collected through various sources. Of course, these types of data characteristics often include unstructured characteristics compared to data collected in other ways [10].

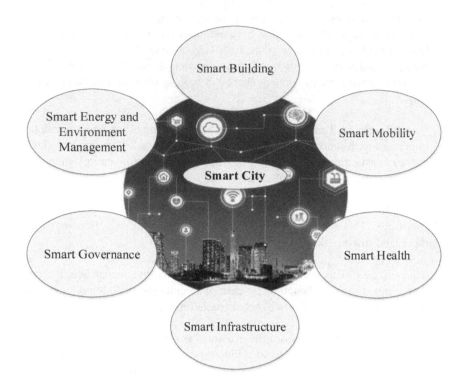

**Figure 7.2**   Various key components of a typical smart city

Big data and cloud computing technology involved in the smart city so that the sensor devices embedded in various smart applications. Besides this, other devices are connected to the cloud computing infrastructure that exchanges information to create large numbers of structured data. This large amount of unsolicited data is collected and stored in the cloud. Therefore, the programming model to process large data sets with parallel algorithms can be used for data analysis to obtain values from the stored data.

In order to handle these enormous amounts of data, the ML approach can be a good choice because ML can work with high-volume data reliably and without exhaustion [1, 19]. The major challenges associated with these tasks need efficient and relevant feature extraction from the data. The surge of data and its type and quantity for all the tasks are not the same and therefore the implementation of ML requires specialized algorithms for deploying in different types of data.

The big data can contain leveled data, unleveled data, or a mixture of both. Based on the type and availability of data, network researchers have to train an algorithm with a specific learning model [2]. In the supervised way of learning, the algorithm learns with labeled data [4, 27]. More specifically, in the supervised learning model, both the input and the output are provided. Based on the leveled data, the algorithm predicts future data. In contrast, the unsupervised model [15] is when an algorithm is provided only the input data, without corresponding output data, as a training dataset. The unsupervised algorithms would be able to function freely to learn more from the data and explore the findings as there present no output values or the teachers. On the other hand, the semi-supervised model learns with the combination of both leveled and unleveled data. Typically, this combination contains a small amount of labeled data and a large amount of unlabeled data.

The rest of the chapter is organized as follows. Machine learning frameworks are described in Section 2. Section 3 discusses ML approaches to solve problems. Section 4 presents the smart city infrastructure. The key challenges of designing a smart city are highlighted in Section 5. The implications of machine learning models in the design of smart cities are illustrated in Section 6 and a brief conclusion in Section 7 ends the chapter.

## 7.2   MACHINE LEARNING FRAMEWORKS

The question may arise about the principle of operation of machine learning (ML) approaches for SC management [6]. A good understanding of the ML technique is required for the efficient design and implementation of SC. This section describes the general ML approaches and different ML algorithms with their applications to design a smart city.

### 7.2.1   MACHINE LEARNING APPROACHES

One of the simple approaches in ML algorithm is to train it from the leveled or un-leveled data set to produce a model. New input data is introduced to a ML algorithm and it makes a prediction based on the model. These approaches are evaluated for accuracy and when the system reaches an acceptable level of accuracy, it is treated as a successful model [29]. On the other hand, if the desired response is not obtained from the system, then it is trained again and again by feeding more input data until the response of the system reaches a satisfactory level. Figure 7.3 shows a simple process of ML techniques. However, there are various factors and steps involved in producing a successful model.

### 7.2.2   MACHINE LEARNING MODELS

To work independently with the big sensor data, machines need to be learned from the various leveled or unleveled and different structures of data by engaging mathematical and statistical approaches. These ML models are consisting of four separate families of techniques that are briefly discussed next.

Supervised Learning
Unsupervised Learning
Semi-supervised Learning
Reinforcement Learning

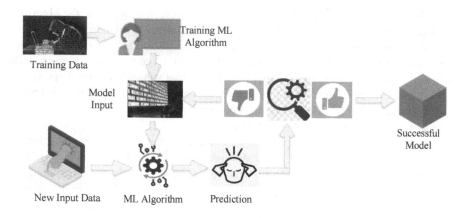

**Figure 7.3**   A pictorial representation of machine learning process

In this section, machine learning purposes are being explored and when and where the specific techniques are adopted are also discussed with proper examples. Some of the specific algorithms are also considered in the discussion.

### 7.2.2.1   Supervised Learning Models

The key idea of a supervised model is that a supervisor is engaged to train a machine like a caregiver/teacher to teach a child to recognize colors, fruits, and numbers. In this method, the teacher checks every step of the child and the child has to learn from the results he/she produces [14].

In supervised ML models, there are two variables, namely $x$ and $y$. Let us suppose, $x$ be the input and $y$ is the output; we can write a function like $y = f(x)$. The goal is to do a mapping so well that whenever it has a new input data x it could predict the output of $y$.

In other words, supervised ML methods, where each instance of a training data set is composed of different input attributes and an extracted output. The input attributes of the training data set can be any kind of data set. For example, it can be a pixel of an image, value of a database, or even be a frequency histogram. For each input instant, an extracted output level is associated. The value can be discreet (presenting a category) or can be a real continuous value. In either case, the algorithm learns the input patterns that generate the extracted output. Now, once the algorithm is trained, it can be used to predict the correct output of a never seen input.

To clarify the above statement, let us consider an example. Figure 7.4 shows a model training for a supervised learning system. One can observe how a model can be developed from the raw input (feeding) to algorithm (apply). For example, a supervisor keeps on correcting the machine or keeps on telling the machine that this is an apple. Hence, this process keeps repeating until we get a final train model. Once the model is ready, it can easily predict the correct output. For instance, after giving

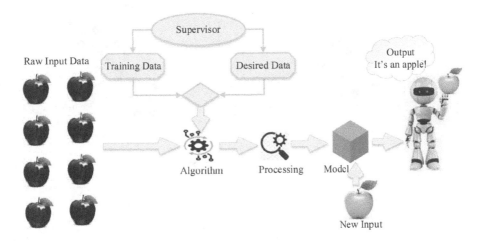

**Figure 7.4**   Model training for supervised learning

a green apple as an input to the machine, and the machine can easily identify it and gives the correct output/result (Figure 7.4).

Some applications of supervised learning applications are given below:

> Speech Automation: Speech automation software like "Cortana" trains using a voice. Once trained, it starts working based on training. This is one application of supervised learning. If someone says "Hey Google, Call Rocky" it automatically calls Rocky from phones detecting the voice note.
>
> Weather Forecast: Considering some prior knowledge like when it is sunny, the temperature is higher and when it is cloudy, the humidity is higher. It can predict the parameters in a given time.
>
> Biometric Attendance: After training the machine with a couple of inputs of biometric identity, once trained, the machine can identify the future input of any specific person.
>
> Bank Sectors: In the banking sector, supervised learning is used to predict the creditworthiness of a credit card holder by building a machine learning model to look for a faulty attribute like providing data on delinquent and non-delinquent customers.
>
> Healthcare Sector: In the healthcare center it is used to predict the patient readmission rate by building a regression model by providing data on the patient treatment administration and readmission to show variables that best correlate with the admission.
>
> Retail Shops: In the retail sector, it is used to analyze the products that customers bye together. It does this by building a supervised model to identify frequent item sets and association rules from transactional data.

Some of the main supervised learning algorithms are discussed next.

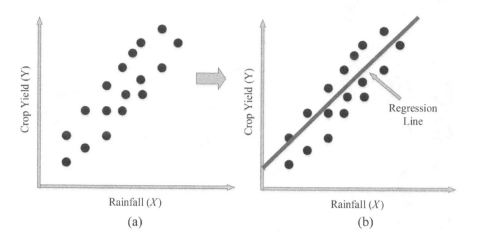

**Figure 7.5**  Simple linear regression model

**Linear Regression:** Linear regression is a statistical model used to predict the relationship between independent and dependent variables by examining two factors. The first one is to examine which variables, in particular, are significant predictors of the outcomes variables, and the second important judgment is how significant is the regression line to make predictions with the highest possible accuracy.

As linear regression is drawing a line between variables, the simplest form of a simple linear regression equation with one dependent and one independent variable represented by:

$$y = m * x + c \qquad (7.1)$$

where,
$y$ is the dependent variable
$x$ is the independent variable
$m$ is the slope of the line; $m = \frac{(y_2 - y_1)}{(x_2 - x_1)}$
$c$ is the coefficient of the line

Suppose we are going to plot the amount of crop yield based on the amount of rainfall. As we cannot change the rainfall, so this is considered as independent variable $x$. On the other hand, the amount of crop growing is dependent on rainfall, hence, crop yield is considered as dependent variable $y$. We can take this and draw a line in the middle of data, we find the regression line that is seen in Figure 7.5 where we can easily predict the amount of crop production with the amount of rainfall.

If there are more than one independent variables, then the equation for multiple linear regression changes to:

$$y = m_1 * x_1 + m_2 * x_2 + \dots + m_n * x_n + c \qquad (7.2)$$

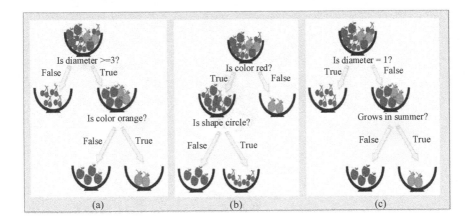

**Figure 7.6** Random decision tree

where, $x_1$, $x_2$, $x_3$ are the independent variables and, $m_1$, $m_2$, $m_3$ are their corresponding slops with still single coefficient $c$.

Some of the applications for linear regression in ML is given below:

The economic growth of a country is predicted by calculating GDP by linear regression algorithm.

This method can be used to predict the price of a product in the future.

Players' scores can be predicted by evaluating the performance of previous matches.

**Random Forest:** Random forest or random decision forest is a method that operates by constructing multiple decision trees during training phases. The decision of the majority of the tree is chosen by the random forest as the final decision. A decision tree is a tree-shaped diagram used to determine a course of action. Each branch of the tree represents a possible decision, occurrence, or reaction.

Figure 7.6 shows the process of a random decision forest work. Starting from Figure 7.6 (a), where a bowl of fruits is compared with a diameter and divided into two parts. Here, the cherries are separated. The lemons and apples have stayed together. In the last step, it is checked whether it's color is orange or not. Finally, the three types of fruits are separated from one another. Similarly, color and shape are compared in Figure 7.6 (b), whereas the diameter and the grown season is checked in Figure 7.6 (c).

With the presence of some missing data, random forest works also well. Consider the image of Figure 7.7 where its color is missing. This unknown object also can be identified by applying a decision tree shown in Figure 7.6. The unknown object has a diameter of 3 but it's color is not sure. However, it goes to identify as orange and it's shape is round (Figure 7.6 (a)). In the next algorithm (Figure 7.6 (b)), if we consider its color is red and as it is circle

Information Given:

Diameter = 3
Grows in Summer = Yes
Shape = Circle

**Figure 7.7**  Detection object with missing data

than it is identified as cherry. Figure 7.6 (c) as it has diameter 3 and grows in summer, it definitely is identified as an orange. Finally, if we sum up these three results, two decisions are voted for an orange (Figure 7.6 (a) and Figure 7.6 (c)) and one decision is for cherry (Figure 7.6 (b)). Therefore, orange is voted more than cherry, hence, it is said that the object is an orange.

Less training time is required for the random forest algorithm. The use of multiple trees reduces the risk of over-fitting. It runs efficiently on a large database and produces highly accurate predictions. It can also predict the accuracy of data with the presence of missing data.

The applications of the random forest are given as follow:

The random forest can be used in remote sensing devices, for example, the uses in ETM devices provide accurate images with high accuracy on the earth's surface. The training time is comparatively less.

It performs well in object detection specially traffics on a road. Multiclass object detection is done using a random forest algorithm in a complicated environment.

This is used in a game console called Kinect. It can also track body movements and recreates it in the game.

**Support Vector Machines:** The support vector machine (SVM) is a supervised learning method that looks at data and sorts it into one of the two categories. In SVM, we have taken some leveled sampled data of apple and strawberry. Then we draw a line between the two groups. This split allows us to take new data and place it in the appropriate group. Now one can predict the unknown object that as shown in Figure 7.8. At the same time and with the same model, the system allows high dimensional input space which is indeed one of the main advantages of SVM.

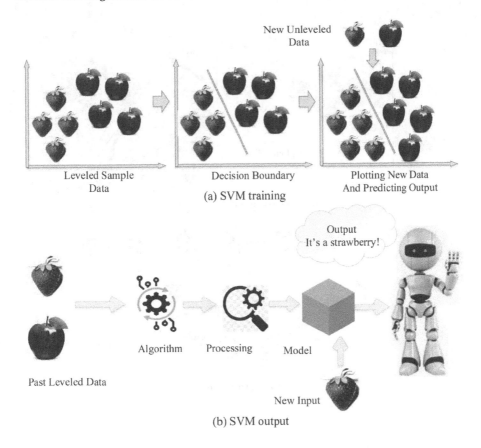

New Unleveled Data

Leveled Sample Data

Decision Boundary

Plotting New Data And Predicting Output

(a) SVM training

Output It's a strawberry!

Past Leveled Data

Algorithm    Processing    Model

New Input

(b) SVM output

**Figure 7.8**   Support vector machine algorithm

Now the question arises about the best position to fit the hyperplane. This is also known as a decision boundary because it provides the correct decision where should the new coming data best fits. The appropriate positioning of hyperplane is discussed in Figure 7.9. The first step to draw the hyperplane is to find the shortest distance between the two clusters and the points of the nearest two cluster is called support vector. Figure 7.9(a) shows the linearly separated data. Now the next step is to find out the maximum distance between the two support-vector. In this case, the distance of $m_1$ is smaller than the distance $m_2$, therefore, the latter would be the best choice of considering the hyperplane.

However, if the data is aligned in a 1-D plane as shown in Figure 7.9(b) then the previous approach is not suitable. In this case, it is necessary to convert the 1-D view to the 2-D view. For the transformation, we use the Kernel function which takes one-dimensional input and converts it into two-dimensional output. Therefore, the 3-D conversion makes it easy to draw a line between the two clusters shown in Figure 7.9 (b).

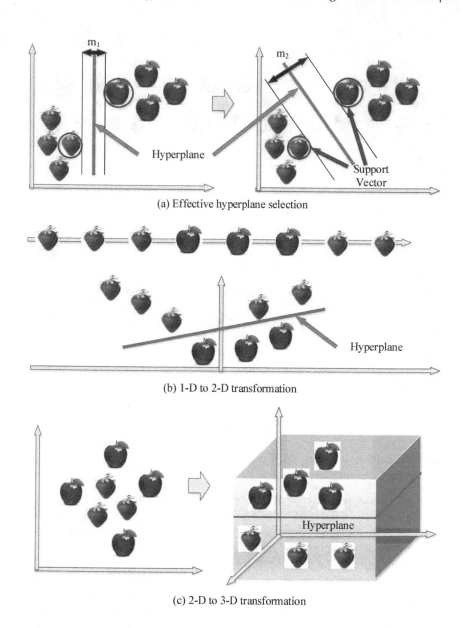

(a) Effective hyperplane selection

(b) 1-D to 2-D transformation

(c) 2-D to 3-D transformation

**Figure 7.9**   Selection of hyperplane of SVM algorithm

On the other hand, if the data is not linear and mix together then the scenario becomes more complicated as shown in Figure 7.9 (c). Here also the Kernel function is applied to convert the two-dimensional data into a three-dimensional array. After this conversion, we can place the hyperplane that can split the data as shown in Figure 7.9(c).

This can also work with sparse document vectors. Finally, the regularization parameter helps us to figure out whether we have biased or over-fitting of data. In SVM, it naturally avoids the over-fitting and biased problem of data that suffers many of the other algorithms. The applications of SVM in smart grid can be found in [17, 18]. The SVM has also been extensively used for other applications such as listed below.

SVM is highly used in face detection.
Used in text and hypertext categorization.
Accurate outputs are obtained in the classification of images and bioinformatics.

### 7.2.2.2 Unsupervised Learning Models

The unsupervised learning models derived from supervised learning with no supervisor to train the model and also there is no predefined correct answer. Various algorithms have been developed to obtain similar types of data and to present the interesting structure in the data. Mathematically, unsupervised machine learning only involves input data x and no corresponding output variable. The goal of unsupervised machine learning is to model the underlining structure or distribution in data in order to learn more about data [27].

In the unsupervised learning approach, data instant of a training data set does not have expected output; the unsupervised algorithm detects pattern based on the unique characteristic of input data. An example of a machine learning task that applies unsupervised learning is clustering. In this task, similar data instances are group together to identify clusters of data.

Figure 7.10 shows the unsupervised learning model. One can observe that initially a variety of fruits are considered as input. Let us consider $x$ be the sets of fruits, the model will create clusters based on its strain using an unsupervised learning algorithm. It will group similar fruits to make a cluster.

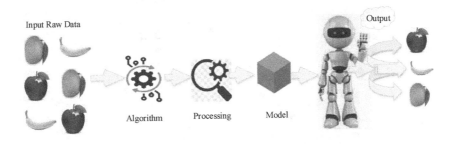

**Figure 7.10**   Model training for unsupervised learning

The unsupervised learning algorithm will process an unleveled data set and based on its characteristics it will group the images of three different clusters of data.

Despite the ability of grouping similar data into the clusters, the algorithm is not capable to add levels to the group. This unsupervised learning algorithm knows only about data instances that are similar. However, this class of algorithm cannot identify the meaning of the group.

The applications of unsupervised machine learning are given below:

Banking Sector: In this scenario, the unsupervised machine learning is used to segment customers using behavioral characteristics to develop multiple segments through clustering.

Healthcare Sector: In the healthcare sector, unsupervised machine learning can be used to categorize MRI data using either normal or abnormal images. The system uses unsupervised learning techniques to build a model that learns from different features of images to recognize a pattern.

Retail Sector: The unsupervised machine learning approach can be used in the retail sector to help customers to find the right products based on customers' past purchasing history. The system does this by building a collaborative filtering model based on the history of past purchasing.

There are various unsupervised learning algorithms are reported in the literature. We briefly discuss some of the most popular algorithms next.

K-Means Algorithm: K-means performs the division of objects into clusters that are similar between them and are dissimilar to the objects belonging to another cluster. If we have some unleveled data, we can cluster it with this algorithm. The term K is a number, therefore, we need to tell the system how many clusters are needed to perform.

To get an insight into this algorithm, let us consider an example of a cricket match. The batsman ran more but get less wickets. In contrast, the bowlers run less but get the wicket most. In Figure 7.11(a), we plot runs against wickets. Figure 7.11(b) shows the clustering model of Figure 7.11(a).

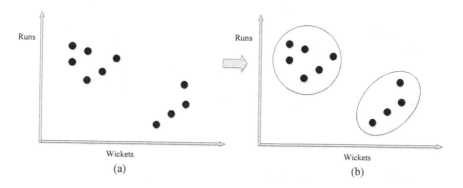

**Figure 7.11**   Runs vs. wickets

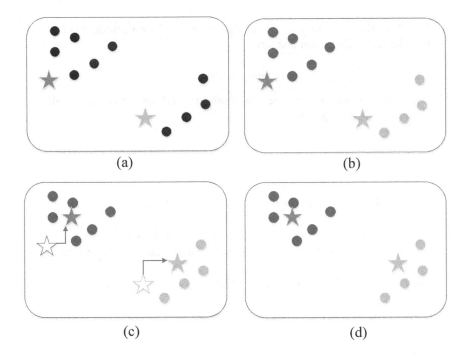

(a)          (b)

(c)          (d)

**Figure 7.12** Relocation of the centroids

In the first step, we need to select the number of clusters that we want to identify in the dataset. This is the "K" in K-means clustering. For this case, let us consider, K = 2, and take two random points on the graph (Figure 7.12(a)). We then determine the distance of each of the data points from each of the randomly assigned points. The points with less distance are assigned for the one type of data that is represented in Figure 7.12(b).

The next step is to determine the actual center point of the actual center points of these two clusters. This process is continued unlit the relocation of the centroid is getting fixed as shown in Figure 7.12(c), and the final cluster is shown in Figure 7.12(d). It is found that the algorithm is converged.

Assuming we have inputs $(x_1, x_2, x_3,....)$ and we are splitting it into $k$ clusters. The relocation process of centroid can be divided into four steps which are described below:

*Step 1:* The first step is to pick $k$ random points as cluster centers that called centroids. Let's assume these are $c_1, c_2, c_3,...c_k$ and we can say that

$$C = c_1, c_2, c_3, ..., c_k \qquad (7.3)$$

Where, $C$ is the set of all centroids.

*Step 2:* In this step, we assign each data point to closest center, this is done by calculating Euclidean distance

$$arg\ min_{c_i \in C}\ dist\ (c_i, x)^2 \qquad (7.4)$$

*Step 3:* In this step, we find the new centroid by taking the average of all the points assigned to that cluster.

$$c_i = \frac{1}{|S_i|} \sum_{x_i \in S_i} x_i \qquad (7.5)$$

where, $S_i$ is the set of all points assigned to the $i$ th cluster.

*Step 4:* In this step, we repeat step 2 and step 3 until none of the cluster assignments change that means until our clusters stable, we repeat the algorithm.

The practical applications of the K-Means algorithms include

Grouping students based on their grades/scores

The search result is grouped to form clusters

Wireless sensor networks where the clustering algorithms play a big role in finding a cluster head that collects all the data of the respective cluster.

Hierarchical Clustering Algorithm: The method of dividing the objects into clusters that are similar between them and are dissimilar to the objects belonging to another cluster. Hierarchical clustering is separating data into different groups based on some measure of similarity. There are two types of Hierarchical clustering; one is "Agglomerative" which brings things together and the other one is "Divisive" which divides things apart. The Agglomerative clustering follows a bottom-up approach whereas Divisive clustering is a top-down approach (Figure 7.13).

Suppose we have few points on a plane as shown in Figure 7.14(a) and we try to find the least distance between two data points to form a cluster. After finding the least distance between two points we need to form clusters of multiple points. This is represented in a tree-like structure called Dendrogram. For three sets of clusters shown in Figure 7.14(a), the corresponding Dendrogram figure is shown in Figure 7.14(b). Then for two nearest points, again we make one dendrogram. The process ends when we make one cluster.

Some applications of Hierarchical clustering are given below:

Widely uses in customer segmentation in super shops.

It also uses social network analysis to short what are the items like or dislike the customers more.

For city planning, this algorithm plays a vital role to separate different zone with one another such as industrial, commercial, residential.

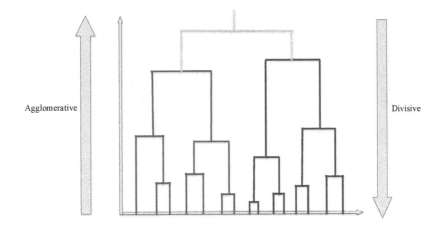

**Figure 7.13**   Agglomerative and divisive algorithms

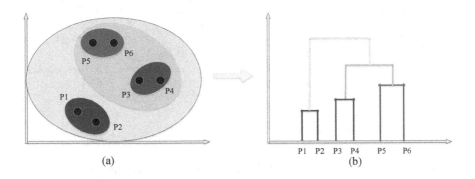

**Figure 7.14**   Hierarchical clustering algorithm

### 7.2.2.3   Semi-Supervised Learning Models

The semi-supervised learning model is a combination of supervised and unsupervised learning models. However, in semi-supervised learning models, datasets are divided into two parts: a labeled part and an unlabeled part. This technique is often used when labeling data or collecting tagged data is very difficult or expensive. The portion of tagged data may also be of low quality [25]. For example, if we do medical imaging for cancer, keeping doctors tagged in the datasets is very expensive to work on and also these doctors need to do more urgent work. The doctor has tagged one part of the dataset and left the other untagged. Finally, this machine learning technique has been shown to work well with good precision, even if the dataset is partially labeled. Additionally, unlabeled data is easily accessible and also available in large quantities.

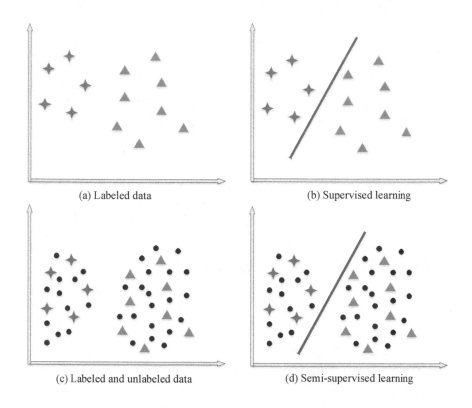

(a) Labeled data          (b) Supervised learning

(c) Labeled and unlabeled data          (d) Semi-supervised learning

**Figure 7.15**   Learning data set of semi-supervised models

Figure 7.15(a) shows some labeled data and supervised learning involves finding a boundary between different types of labeled data (Figure 7.15(b)). On the contrary, for the case of semi-supervised learning, there are both labeled and unlabeled data (Figure 7.15(c)). But the amount of unleveled data is quite high than the leveled data. Training is performed in such a way that the model could divide them into similar parts (Figure 7.15(d)). In this case, the division is done taking the consideration of leveled data with the availability of unlabeled data.

A popular semi-supervised learning algorithm is described below:

> Generative Adversarial Networks: A common training method used in medical software diagnostic with a very small set of labeled data is Generative Adversarial Networks (GAN) which is a type of semi-supervised learning method. From Figure 7.16, it is seen that real samples contain real images that are mixed with some latent space which generates noise. This data is fed into the Generator that generates some fake samples with fake images. Both the real sample and the output of the generator are mixed and compared in Discriminator. The Discriminator passes all the images and fine-tunes and

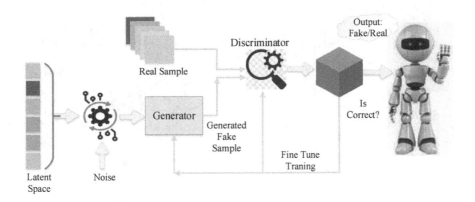

**Figure 7.16**   Model training for GAN

gets trained. After that, it is identified whether the image is real or fake. By this method, the GAN is used to identify the real images. The GAN is very efficient as it reduces the time that compared to unsupervised learning modeling of identifying images from the fake images.

### 7.2.2.4  Reinforcement Learning Model

Reinforcement learning is a class of ML algorithm which allows the software agents and machines to automatically determine the ideal behavior with a specific context to maximize its performance. These learning models interact with both the environments and the learning agents. The learning agent leverages two mechanisms namely, explantation and exploitation. When learning agent acts as "trial and error" basis it is termed as exploration, and when it acts on the knowledge gained based on the environment it is referred to as exploitation. The environment rewards the agent for the correct actions and the agent improves the environment knowledge to select the next action [9].

In Figure 7.17, if the machine is confused about whether the fruit is an apple or not; then the machine is trained using a reinforcement learning algorithm. If the system makes the correct decision it gets a reward point for it and in the case of a wrong decision, it gets a penalty. Once the training is done, the machine can easily identify which one of them is an apple.

The applications for reinforcement machine learning algorithms are given below.

Banking Sector: In the banking sector, reinforcement learning is used to create the next best offer model for a call center by building a predictive model that learns over time as users accept and reject offers made by the sales staff. Healthcare Sector: In the healthcare sector, it is used to allocate the scars medical resources to handle different types of ER cases by building a mark of the decision process that learns treatment strategy for each type of ER cases.

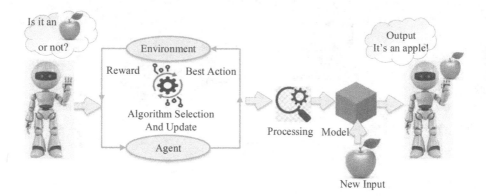

**Figure 7.17**    Model training for reinforcement learning

Retail Sectors: In the retail sector it can be used to reduce assess stock with dynamic pricing. By building a dynamic pricing model that adjusts the price based on customers' response to the offers.

## 7.3    PROBLEM-SOLVING USING MACHINE LEARNING TECHNIQUES

The problems of smart cities (SC) can be categorized into five groups as shown in Figure 7.18. The details of each problem are given below:

**Classification Algorithms:** The classification algorithm is used when we set the number of outputs. For example, "Is it cold?". The answer is either "yes or no" or the question like, "Will you go to a party today?" The answer comes "yes, no, or maybe". For a particular case, when an answer is known to us and if it could not be anything other than that, then we use a

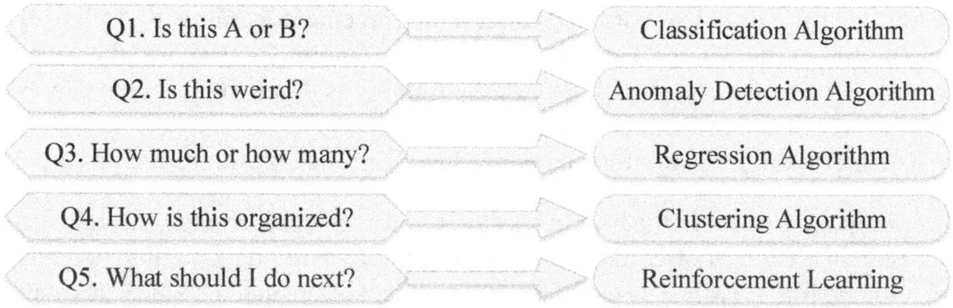

| | |
|---|---|
| Q1. Is this A or B? | Classification Algorithm |
| Q2. Is this weird? | Anomaly Detection Algorithm |
| Q3. How much or how many? | Regression Algorithm |
| Q4. How is this organized? | Clustering Algorithm |
| Q5. What should I do next? | Reinforcement Learning |

**Figure 7.18**    Problem-solving using machine learning techniques

Anomaly

**Figure 7.19**   Anomaly detection

classification algorithm. When we have only two choices (yes or no), it is called 2 class classification. When we have more than two choices (yes, no, or maybe) than it is called multi-class classifications.

Anomaly Detection Algorithms: In this algorithm, we use certain patterns and we get notified when there is some anomaly or any disorder or not usually happens. In Figure 7.19, there is a pattern of man, suddenly a red man comes up. In the system, it is considered as an anomaly as it breaks the pattern and this is not expected usually.

In real life, credit card companies use this anomaly detection algorithm and flag any transactions which are not usually as per transaction history.

**Regression Algorithms:** Whenever the expected outcome is come up with value than we use the regression algorithm. The question like "What will be the temperature tomorrow?" The answer should be a value by analyzing mathematical equations. Similarly, the question like "How much discount can I give to a particular item?" The answer also is a number.

**Clustering Algorithms:** This algorithm is used to establish a structure from the unstructured data set. Suppose we have mixed unstructured data as shown in Figure 7.20, if it is passed through the clustering algorithm then we find some groups of pattern data. Hence, the data will be categorized in groups A, B, C, and D. However, the computer does not understand anything but it can only filter the similar patterns of data.

Reinforcement Algorithms: These algorithms are designed as to how brains of humans respond to punishments and rewards that they learn from outcomes and decide on the next action. Whenever machines need to provide a decision or come up with solutions than this algorithm is used considering the past event feedback. For example, if someone trains a computer how to play chess than it should be used by a reinforcement algorithm. When it has learned or when the model is created for that and the game is played by the computers each decision the computers make is taken form the reinforcement algorithm.

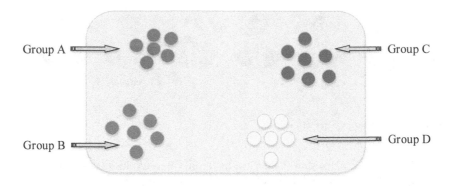

**Figure 7.20**   Clustering algorithm

## 7.4   SMART CITY DESIGN INFRASTRUCTURE

This section briefly describes the application of ML for smart cities. The concept of ML applications in smart cities is illustrated in Figure 7.21. One can observe that this framework consisting of three levels, namely, IoT devices, fog computing, and cloud computing [22]. The devices/sensors of the smart city are controlled by intelligent software that is correlated with the fog or cloud. Raw data that is generated from various sensor devices in a typical smart city can be sent to the fog and then to the cloud for storage and processing for performance prediction. For example, street lights can automatically be turned off or turned on based on weather conditions.

The inspiration behind this architecture is that as data travels through the smart city infrastructure, images from the deeper levels of data abstraction and knowledge presentation can be encapsulated. At the highest level, city-wide abstraction is required to manage the city's resources and services over the long term. At the lowest level, the data generated by sensors and smart objects are used to manage resources and services in the short term. Additionally, fog-based analytics support local activities in predefined contexts, but cloud-based analytics can cover larger geographic regions in different contexts.

The level of IoT infrastructure where sensors and devices with limited resources perceive the environment. The resource limitations of these devices make it difficult to implement large and complex learning models. Instead, multiple shallow machine learning methods can be applied to make them smarter in the context of devices, including ineffective and semi-supervised methods (for example, closest K neighbors, support vector machines, etc.). However, modern and advanced learning models, such as deep learning, should be used to bring analysis and intelligence closer to data sources (for example, end-users, IoT devices with limited resources). A postgraduate research pathway transcends resource limitations to allow deep neural network models of these devices to use them. In recent years, various methods have been proposed to compress or cut deep neural networks so that they can be loaded into devices

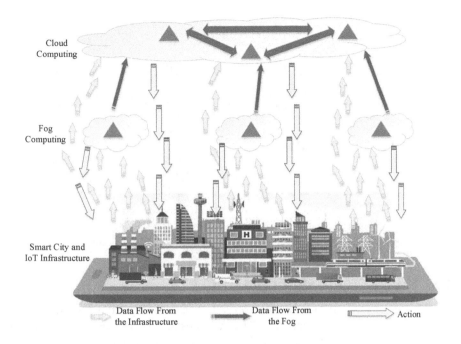

Cloud
Computing

Fog
Computing

Smart City and
IoT Infrastructure

Data Flow From
the Infrastructure

Data Flow From
the Fog

Action

**Figure 7.21**   Machine learning applications for smart cities

with limited IoT resources, portable electronic devices, and smartphones [10]. Using such compressed neural networks, it is possible to integrate deep reinforcement learning with these devices. At the fog computing level, the raw data is collected and sent to the cloud computing level. Narrow deep learning models, DRLs, and semi-supervised methods can be used at this level because resources at this level have fewer limitations than IoT organizations [16]. The proposed semi-supervised DRL method is also applicable at this level. Also at this level, light intelligence must be brought to IoT gateways and proxy servers to enable efficient horizontal integration of services in support of smart city applications.

At the cloud computing level, more complex, large-scale machine learning and data mining frameworks and algorithms can be integrated with semantic learning and ontologies to remove high-level information and patterns from collected data. Deep learning models fit extremely well at this level because they can provide deep data abstraction. Recent advancements in graphics processing unit (GPU) technology, as well as the development of efficient neural network parameter initialization algorithms (e.g. autoencoders), the use of modified linear units (releases), and long-term memory. Therefore, the system allows the perception of efficient models of deeper learning.

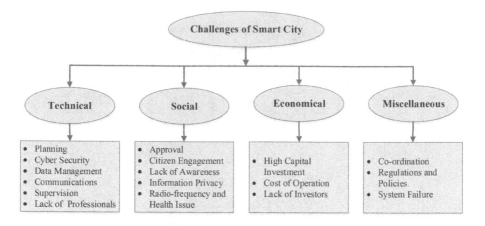

**Figure 7.22**   Challenges in the design and implementation of a typical smart city

## 7.5   SMART CITY DESIGN CHALLENGES

The design of a smart city involves several stages starting from sensors and IoT devices and up to cloud computing. The main obstacle to designing a smart city is to handle a huge amount of data that is produced from different IoT devices in real-time. Effective control and management of big data is important for an accurate and reliable system. The major challenges to designing a smart city are shown in Figure 7.22 and some of them are discussed below.

### 7.5.1   TECHNICAL CHALLENGES

Planning: To design a smart city, proper and appropriate planning is considered the most vital tasks. It also involves the infrastructural and architectural design of a city. By allocating space for all equipment and considering the other necessary factors in mind, we can overcome this problem.

Cyber-security: Another important challenge for smart cities is cyber-security. If the hack has occurred, it becomes a matter of disaster for the whole city as the system is interlinked with one another. There are some common cyber threats are hackers, zero-day, malware, etc. These challenges are not easy to handle if they are applied to the system. The cyber threats are not limited to a certain type of number therefore, it is required to build-up higher technology to provide good protection against these threats.

Data Management and Communications: Different types of meters, sensors, and controllers are used in various stages to communicate with one another in a smart city and they produce in real-time. That data needs to handle effectively for compiling and predict the decision. Otherwise, the system significantly slows down to handle the Big data and it may lose stability.

Supervision: Continuous monitoring and supervision are necessary for every step for the successful operation of the smart city. A failure of one stage may shut down the whole network. This sometimes becomes challenging due to a lack of IT professionals. However, this should be overcome by engaging more manpower in every stage.

## 7.5.2  SOCIAL CHALLENGES

Approval: Approval of the smart city project plan from Government is one of the major challenges as it required lots of considerations. The governing body may not be motivated to convert conventional cities to smart cities as it needs a long time project plan and a high volume of capital investment.

Citizen Engagement: The citizen may be discouraged with the project as they do not know the benefits of a smart city. However, they can be motivated with proper counseling and providing accurate information.

Lack of Awareness: As the smart city is a new concept, many citizens may raise their voice against the smart city project as they are happy with their existing conventional city. Therefore, policymakers also need to consider this scenario and clarify the outcomes to the citizen. Additionally, some flexible offers for customers should be considered.

Information Privacy: Privacy is important for both Government and citizen. This IoT system is highly porn to cyber-attack and within a second huge amount of data can be stolen. These data can also be possible to do modifications. Therefore, to maintain the faith of citizens, all kinds of threats should be protected.

Radiofrequency (RF) signal and health issue: Some health-conscious people and medical experts claim that IoT devices emit high-frequency RF signals that can cause health problem in long term. This might be an obstacle to run the smart city project.

## 7.5.3  ECONOMIC CHALLENGES

High Capital Investment: The smart city project requires a high capital investment initially that may be a barrier to implement the project. Although, it can overcome with a long-term plan by applying a new but small amount of service charges to the citizen when they use different software related to this. In this way, both the government and citizens are benefited.

Lack of Investor: Generally, as the smart city project is a government project, hence, there is less number of investors are found. The government should encourage more investors to invest money with this project with a lucrative profit.

Cost of Operation: As a smart city concept involves several equipment and steps, it requires maintenance on a regular basis. Therefore, IT experts and professionals are required to run the project successfully.

### 7.5.4   MISCELLANEOUS CHALLENGES

Co-ordination: The mega project like smart city, where approximately all major engineering background people are involved with non-technical people like investors, policymakers, etc., therefore, sometimes coordination may become difficult and this may lead a serious problem.

Regulations and Policies: From starting to end, regulations and policies should be analyzed and revised if needed. The government should keep the citizen in most priority to make a new rule. The active participation of citizens is expected for a successful project.

System Failure: In case of any system failure, a backup plan is needed to be prepared and research is needed to restore the network in a fast process to maintain the continuity of the system.

## 7.6   IMPLICATIONS OF ML MODELS IN THE DESIGN OF SMART CITIES

The concept of SC is introduced at the beginning of the 20th century and since then it opens up a lot of opportunities for network researchers for the design and implementation of SC. In fact, SC is the combination of smart power, smart health, smart mobility, in a word SC is essential for our smart life. In the successful operation and management of SC, in every stage, IoT devices and technology are used to communicate with each device. The parameters of each device are unique. Therefore, to combine all these things in the same platform is a matter of challenge. Some open research problems are highlighted below.

Real-time monitoring is the key feature of SC, hence, the number of data increases dramatically. Additionally, the overall system may also slow down. Therefore, handling big data becomes a matter of challenge, and research is needed to overcome this situation.

New communication infrastructure is needed to develop to meet up the requirements of SC and to enhance reliability and sustainability.

Information security is one of the biggest challenges as the IoT structure is a cyber-physical system and highly porn to cyber threats. The data can be modified and stolen if proper security is not developed and research is needed to make a robust multi-layer protection scheme.

To make a city smarter, ML approaches are considered to predict unknown objects or behaviors. As the application fields and objectives are varied from one another, research is needed to develop the fast and accurate algorithm of ML.

## 7.7   CONCLUDING REMARKS

There are various machine learning (ML) models and algorithms that can be used to learn from sensor data collected through smart cities (SC) infrastructure. In this chapter, we presented a number of ML models to understand the practical implications of these models suitable for designing and promoting sustainable SC. These computational ML models and algorithms are necessary to assist the process of building SC as they are fundamental to the decision making for improved SC services to citizens. We highlighted various challenges that arise when utilizing ML to realize SC services. These challenges include technical, social, and economical. We found planning and cybersecurity aspects are the most important technical challenges. This is because appropriate planning is needed to design a smart city that involves the infrastructural and architectural design of a city. To mitigate cyber threats it is important to build a technological solution against these threats. The discussion presented in this chapter provides some insights into the use of ML in the design of SC that can help network researchers and engineers to contribute further towards developing SC. Future research work could include analyzing the ML techniques to create a robust model to perform enhanced predictive functions for SC.

## REFERENCES

1. Omar Y Al-Jarrah, Paul D Yoo, Sami Muhaidat, George K Karagiannidis, and Kamal Taha. Efficient machine learning for big data: A review. *Big Data Research*, 2(3):87–93, 2015.
2. Ahlam Althobaiti, Anish Jindal, and Angelos K Marnerides. Scada-agnostic power modelling for distributed renewable energy sources. In *2020 IEEE 21st International Symposium on" A World of Wireless, Mobile and Multimedia Networks" (WoWMoM)*, pages 379–384. IEEE, 2020.
3. Gagangeet Singh Aujla, Anish Jindal, and Neeraj Kumar. Evaas: Electric vehicle-as-a-service for energy trading in SDN-enabled smart transportation system. *Computer Networks*, 143:247–262, 2018.
4. Gagangeet Singh Aujla and Neeraj Kumar. Mensus: An efficient scheme for energy management with sustainability of cloud data centers in edge–cloud environment. *Future Generation Computer Systems*, 86:1279–1300, 2018.
5. Gagangeet Singh Aujla, Neeraj Kumar, Mukesh Singh, and Albert Y Zomaya. Energy trading with dynamic pricing for electric vehicles in a smart city environment. *Journal of Parallel and Distributed Computing*, 127:169–183, 2019.
6. Gagangeet Singh Aujla, Amritpal Singh, and Neeraj Kumar. Adaptflow: Adaptive flow forwarding scheme for software defined industrial networks. *IEEE Internet of Things Journal*, vol. 7, no. 7, pp. 5843-5851, 2019.
7. Gagangeet Singh Aujla, Maninderpal Singh, Arnab Bose, Neeraj Kumar, Guangjie Han, and Rajkumar Buyya. Blocksdn: Blockchain-as-a-service for software defined networking in smart city applications. *IEEE Network*, 34(2):83–91, 2020.
8. Gagangeet Singh Aujla, Mukesh Singh, Neeraj Kumar, and Albert Zomaya. Stackelberg game for energy-aware resource allocation to sustain data centers using res. *IEEE Transactions on Cloud Computing*, vol. 7, no. 4, pp. 1109-1123, 2017.

9. Ramin Ayanzadeh, Milton Halem, and Tim Finin. Reinforcement quantum annealing: A hybrid quantum learning automata. *Scientific reports*, 10(1):1–11, 2020.

10. Min Chen, Shiwen Mao, and Yunhao Liu. Big data: A survey. *Mobile networks and applications*, 19(2):171–209, 2014.

11. UN DESA. 68% of the world population projected to live in urban areas by 2050, says UN. *United Nations Department of Economic and Social Affairs*, 2018.

12. João P. Gouveia, Júlia Seixas, and George Giannakidis. Smart city energy planning: Integrating data and tools. In *Proceedings of the 25th International Conference Companion on World Wide Web*, pages 345–350. 2016.

13. Ibrahim Abaker Targio Hashem, Victor Chang, Nor Badrul Anuar, Kayode Adewole, Ibrar Yaqoob, Abdullah Gani, Ejaz Ahmed, and Haruna Chiroma. The role of big data in smart city. *International Journal of Information Management*, 36(5):748–758, 2016.

14. Tabasum Mirza1and Malik Mubasher Hassan. Prediction of school drop outs with the help of machine learning algorithms, *GIS Science Journal*, vol. 7, no. 7, pp. 253-263, 2020.

15. Trevor Hastie, Robert Tibshirani, and Jerome Friedman. Unsupervised learning. The elements of statistical learning (pp. 485-585), 2009.

16. Anish Jindal, Gagangeet Singh Aujla, Neeraj Kumar, Radu Prodan, and Mohammad S Obaidat. Drums: Demand response management in a smart city using deep learning and SVR. In *2018 IEEE Global Communications Conference (GLOBECOM)*, pages 1–6. IEEE, 2018.

17. Anish Jindal, Amit Dua, Kuljeet Kaur, Mukesh Singh, Neeraj Kumar, and Sukumar Mishra. Decision tree and SVM-based data analytics for theft detection in smart grid. *IEEE Transactions on Industrial Informatics*, 12(3):1005–1016, 2016.

18. Anish Jindal, Neeraj Kumar, and Mukesh Singh. A data analytical approach using support vector machine for demand response management in smart grid. In *2016 IEEE Power and Energy Society General Meeting (PESGM)*, pages 1–5. IEEE, 2016.

19. Anish Jindal, Alberto Schaeffer-Filho, Angelos K Marnerides, Paul Smith, Andreas Mauthe, and Lisandro Granville. Tackling energy theft in smart grids through data-driven analysis. In *2020 International Conference on Computing, Networking and Communications (ICNC)*, pages 410–414. IEEE, 2020.

20. Athar Ali Khan, Mubashir Husain Rehmani, and Martin Reisslein. Cognitive radio for smart grids: Survey of architectures, spectrum sensing mechanisms, and networking protocols. *IEEE Communications Surveys & Tutorials*, 18(1):860–898, 2015.

21. Neeraj Kumar, Tanya Dhand, Anish Jindal, Gagangeet S. Aujla, Haotong Cao, and Longxiang Yang. An edge-fog computing framework for cloud of things in vehicle to grid environment. In *2020 IEEE 21st International Symposium on A World of Wireless, Mobile and Multimedia Networks" (WoWMoM)*, pages 354–359. IEEE, 2020.

22. Mehdi Mohammadi and Ala Al-Fuqaha. Enabling cognitive smart cities using big data and machine learning: Approaches and challenges. *IEEE Communications Magazine*, 56(2):94–101, 2018.

23. Saeed Nosratabadi, Amir Mosavi, Ramin Keivani, Sina Ardabili, and Farshid Aram. State of the art survey of deep learning and machine learning models for smart cities and urban sustainability. In *International Conference on Global Research and Education*, pages 228–238. Springer, 2019.

24. Saeed Nosratabadi, Amir Mosavi, Shahaboddin Shamshirband, Edmundas Kazimieras Zavadskas, Andry Rakotonirainy, and Kwok Wing Chau. Sustainable business models: A review. *Sustainability*, 11(6):1663, 2019.

25. Y CA Padmanabha Reddy, P Viswanath, and B Eswara Reddy. Semi-supervised learning: A brief review. *International Journal of Engineering and Technology*, 7(1.8):81, 2018.

26. Ahsan Saleem, Abid Khan, Saif Ur Rehman Malik, Haris Pervaiz, Hassan Malik, Masoom Alam, and Anish Jindal. Fesda: Fog-enabled secure data aggregation in smart grid IoT network. *IEEE Internet of Things Journal*, 7(7):6132–6142, 2020.

27. Ramadass Sathya and Annamma Abraham. Comparison of supervised and unsupervised learning algorithms for pattern classification. *International Journal of Advanced Research in Artificial Intelligence*, 2(2):34–38, 2013.

28. Bhagya Nathali Silva, Murad Khan, and Kijun Han. Towards sustainable smart cities: A review of trends, architectures, components, and open challenges in smart cities. *Sustainable Cities and Society*, 38:697–713, 2018.

29. Tony Thomas, Athira P Vijayaraghavan, and Sabu Emmanuel. *Machine Learning Approaches in Cyber Security Analytics*. Springer, 2020.

# 8 Machine Learning Models in Smart Cities – Data-Driven Perspective

*Seyed Mahdi Miraftabzadeh*
Politecnico Di Milano, Italy

*Michela Longo*
Politecnico Di Milano, Italy

*Federica Foiadelli*
Politecnico Di Milano, Italy

## CONTENTS

## 8.1   INTRODUCTION

Nowadays, we hear more and more often about how our cities must evolve to become smart cities. To achieve this, however, it is necessary to collect and analyze an incredible amount of data in real-time. IoT platforms, created in some cases by the same players who installed other infrastructures in our cities, enable data collection; however, it is necessary to analyze this data to transform it into information.

It is essential to set up this type of analysis based on historical data. However, the data cannot be analyzed in such a timely way to make our cities smarter. To do this, it is necessary to rely on machine learning and artificial intelligence systems to be able to use the collected data in real-time to allow actions that help citizens have a better experience within urban centers. To this end, machine learning algorithms have broad applications in smart cities. Figure 8.1 summarizes some of the standard applications briefly.

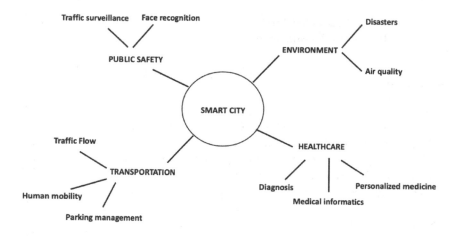

**Figure 8.1**   Machine learning applications in smart cities

Here are some areas where artificial intelligence could be used to improve response times and make our cities smarter:

*Traffic analysis:* using artificial intelligence for image and video recognition applications, together with maps and/or navigation services, or together with the geolocation of smartphones, would allow providing even more precise traffic indications than the current ones do navigation systems. Furthermore, in the event of accidents, the call of emergency vehicles and/or law enforcement agencies could be automated, so as to reduce response times and improve the quality of interventions.

*Intelligent temperature control in buildings:* buildings, especially older ones, are among the primary sources of pollution within urban centers. Using artificial intelligence systems, together with suitable sensors, would allow, with a relatively low cost, to significantly improve the supply of air conditioning inside the buildings, simultaneously reducing the emissions of polluting substances and the costs of air conditioning for citizens, generating thus benefits both individuals and the community.

*Smart parking management:* the use of sensors, together with the geolocation of citizens, could make it easier to identify the free parking spaces in the area of interest, allowing a significant saving of time and a better user experience.

*Sizing of available public transport:* using artificial intelligence applications, through the use of images/videos and the geolocation data of users, allows planning the available means not based on standard planning but based on the actual flows of the passengers within the day.

*Building and infrastructure monitoring:* artificial intelligence can be used to analyze the enormous amount of data that sensors are possibly used to monitor the status of buildings and infrastructures. For example, it can identify potentially dangerous patterns within the data to launch timely alarms in case of problems. The use of intelligent devices then (e.g. valves) could also significantly reduce any damage caused by problems (e.g. water leakage).

These are just a few examples of how artificial intelligence could help us "create" real smart cities or make our cities smarter. Of course, there are potential problems related to the security and privacy of people. For this reason, the legislator must give clear guidelines on how these types of applications must be developed, immediately integrating ethics-by-design logic, then carefully monitoring that the applications manage the data correctly (e.g. correctly anonymize the information or keep it only for the purposes and within the established terms). Suppose there is no absolute certainty about the security of the applications, both from a technical point of view and from regulatory reinforcement, as well as on the impossibility of private subjects to take possession of the personal data relating to individual citizens. In that case, it will not be possible to spread this type of application effectively.

## 8.2   ARTIFICIAL INTELLIGENCE AND THE SMART CITIES

The issues concerning artificial intelligence applications are multiplying both in academic literature and in information in general. The concept of artificial intelligence (the acronym is AI) is not recent but was born in the 1950s, and the British mathematician Alan Turing is generally recognized as the father of computer science and AI. In the years following its development, it suffered several setbacks, thus, the idea of a machine capable of learning (machine learning is also a concept of the past) was abandoned. However, the recent technological revolution has reopened the path to artificial intelligence, thanks above all to novel computing, which gives a computer processing capacity unthinkable until a decade ago, and the availability of huge amounts of data (big data) that can be processed thanks to quantum calculation. This makes AI the potential vector of the most significant transformation capable of entirely changing not only the social, economic, and political structure but of replacing the man-machine interface - already present in many sectors – to the machine-machine interface ( or M2M, from English machine-to-machine), not without causing some doubts on the part of some.

AI programs are already part of our daily lives, even if we still do not use all its capacity. For example, think about the voice assistants or neural networks that use big data to analyze user behavior or even self-driving cars. As it is known, these programs are based on more or less complex algorithms, but it is still an AI defined as "weak" because it is limited to specific areas. However, research on artificial intelligence is experiencing an incredible expansion in all sectors intending to achieve such an AI capable of competing with humans.

The expansion of the Internet of Things (IoT) opens up new possibilities for the evolution in the applications of artificial intelligence, since, if this is currently limited in the field of application, more AI programs interconnected with each other through the 'IoT can lead to the creation of a neuralgic network capable of autonomously managing an entire system, such as urban planning. In the latter case, we speak of smart city, the smart city to which many local administrations of the most developed countries aim to increase automation of municipal services and activities. The implementation and combination of AI, self-learning algorithms (ML, machine learning), and widespread connectivity can collect vast amounts of data within the city, transmit it to a central server, and process it in order to make it easier for urban traffic management, but also be able to operate in the sector of separate waste collection and the health. For the moment the AI, ML, and IoT have found a general application in the traffic management sector, which is increasingly congested in large cities, but the potential for application in the future is broad, mostly due to the growing need to reduce costs and optimize resources available improving the quality of life of citizens, in terms of safety, infrastructure, environment, and sustainability.

Nevertheless, it should be pointed out that a smart city is not a city equipped with IoT technology or only "uses" the internet. The smart city of the future is an integrated network, a "brain" within which they interact, communicating with each other, technology, administration, infrastructure, and citizens. IA and ML allow us to monitor the trends of a city's daily life and then adapt to the needs to make them more efficient and effective. For example, accidents, fires, and disasters of various kinds can also be identified without direct reporting, allowing faster intervention times and in real-time arranging traffic reports to identify alternative routes in order to avoid traffic jams, facilitating circulation, and with effects positive on consumption and $CO_2$ emissions. Similarly, a smart city can offer more excellent security thanks to the technological supports and information immediately available to the police that facilitate their intervention on the territory, including identifying criminals or even those who commit minor crimes and violations of various nature. And again, think of the chronic lack of parking in large cities: the applied artificial intelligence systems can intervene both in the sense of identifying and reporting to motorists the parking spaces available in the area and in the sense of highlighting to the administrations where there is a need to build them of new ones. Finally, also in public lighting, the

AI can find application in order to rationalize its consumption.

The development of AI, ML, and IoT, therefore, seems to be the solution to improve the quality of life in urban centers increasingly congested by traffic, increasingly populous, and with growing management and waste problems. But even in this case, the transformation of the city into a network of interconnections is not without its critical aspects. The main AI applications are based on video surveillance sensors widespread throughout the city center. This involves continuous monitoring of the population and its movements, in addition to the fact that the system collects large amounts of personal data of all kinds, which are then processed according to the type of service for which they are used from time to time (Figure 8.2).

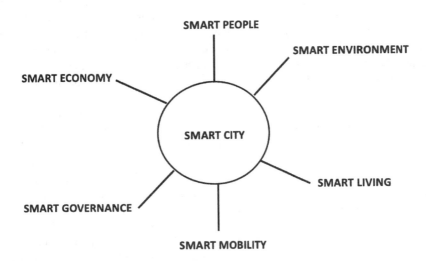

**Figure 8.2** Components of the smart city

## 8.3 MACHINE LEARNING

Machine learning techniques, which were first used in computer systems, are usually applied across engineering and sciences in every field. Furthermore, machine learning methods are broadly used for recommendation, customer behavior prediction, compliance, or risk by commercial sites (e.g. Amazon, eBay), search engines websites Google, recommender systems for streaming services (e.g. Netflix, HBO), financial or banking companies, and advertisers [1]. In simple words, machine learning (ML) implies learning from data without having any explicit knowledge. It is the interdisciplinary research area that incorporates knowledge from different fields, extracts knowledge from experience gathered, solving problems to be used to answer unseen examples of similar questions. Recognizing a cat or dog's image by being shown, many of their images is the most famous machine learning example. In other words, such a model does not follow the step by step instructions on how to differentiate a cat image(s) compare to others. Alternatively, a machine learning model learns what features in the cat images dataset are essential to decide whether a new image contains a cat. This means that a machine learning model can learn from data on how to perform a specific task without having a specific instruction. Therefore, a machine learning performance is strongly related to the volume, variety, and velocity of data, the primary 3 Vs of big data. Machine learning models can detect hidden trends and patterns in datasets, leading to valuable insights and data-driven decision-makers to improve general performance. As it is mentioned, machine learning is the interdisciplinary field between computer science, statistics, mathematic, optimization, artificial intelligence, and domain knowledge. In order to successfully apply this methodology, domain knowledge (understanding the field, applications, or business) of a problem is essential. The domain knowledge helps to understand the meaning of data and their relationships inside the dataset, restrictions, and boundaries, the outcomes, and how to build a meaningful and efficient machine learning model Figure 8.3 [2].

Machine learning is not a new field, however, it is a hot topic because the advanced computing technologies provide accessible means to use its capacity fully. [3]. Big data technologies, introduced as the top topic in 2012 in the high-tech industry, enable machine learning to make accurate predictions or calculated reliable suggestions. Machine learning is broadly used in banking, financial trading, fraud detections, recommendations, data security, personal security, online searching, natural language processing (NLP), self-driving car, and smart cities [4]. In transportation, machine learning is applied to recognize hidden patterns and trends by analyzing various data types such as road traffic during a day. Thus, it helps to make routes more efficient, ease traffic congestions, predict possible problems, and, finally, enhance private and public transportation organizations' profits [3]. In summary, machine learning is a bundle of mathematical and statistical methods that learn to perform a task from data by itself without being explicitly programmed. It is applicable due to the availability of a massive amount of data in different formats, affordable data storage systems, and cheap and powerful computational processing. The techniques used for preparing data for machine learning tasks are the same as those in data mining, predictive analysis, data science, or any other data-driven approaches [5].

## 8.3.1   CATEGORIES OF MACHINE LEARNING TECHNIQUES

Machine learning techniques are generally categorized into supervised, unsupervised, and semi-supervised learning. Supervised learnings refer to problems where the target value or label is known; for example, in recognition of cat images, the target value is the image contains a cat or not. On the other hand, unsupervised learning refers to problems where the target value is unknown. In other words, the output of datasets is not labeled, for example, in cluster analysis, it examines a dataset with

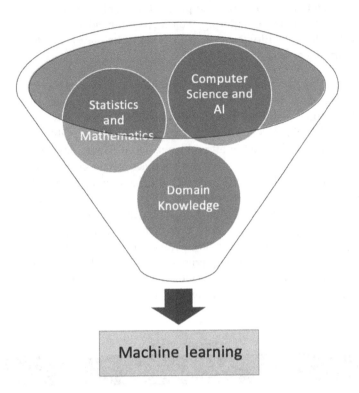

**Figure 8.3**   Machine learning

no pre-existing labels to find undetected patterns to group data. In semi-supervised problems, some data is labeled while the bigger dataset is not [9–11]. The category of machine learning algorithms are listed as follow [6]:

*Classification:* in this supervised kind of problem, the objective is to predict the category of input data.

*Regression:* this important supervised task aims to predict the numeric value.

*Cluster Analysis:* the aim is to group (cluster) a set of similar objects in such a way that they have the most similarities to each other than other groups (clusters) [7].

*Association Analysis:* the goal is to find interesting connections and strict rules to identify associations between items or events in large databases [8].

## 8.3.2   BIG DATA AND MACHINE LEARNING

The massive amounts of data need to proceed for applications, such as drug effectiveness analysis, climate monitoring, website recommendations, etc. Thus, we need to be able to add scalability to machine learning techniques [12]. How do we apply machine learning at scale?

One solution is to scale up by adding more processors, memory, and storage system so as to store and process massive amounts of data. It is also possible to utilize graphical processing units (GPUs) for analyzing datasets in machine learning tasks. Although it seems to be an acceptable approach, this solution cannot be used in all problems, and it is not considered a big data approach. The disadvantages of this approach are the cost of adding specialized hardware such as GPUs and reach the maximum capacity of storage and processing system at some point.

An alternative method, the big data approach, is to scale out. This implies using as many as possible local distributed systems to handle the storage and processing limitations. In this method, the data is spread over systems to attain more power to speed up processing. As it is shown in Figure 8.4, the main idea is to split data into smaller sub-groups, and after processing each sub-group, the results are projected to the merging unit [13].

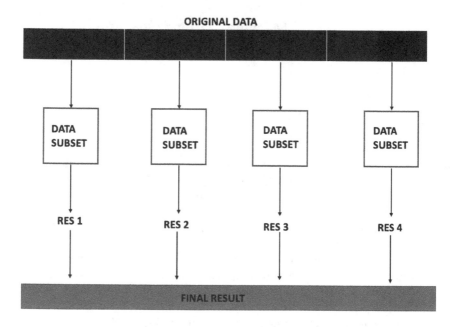

**Figure 8.4**   Divided data into subsets to speed up processing

In the following example, we want to apply the same operation to all the samples in an N samples dataset. In this case, N is four. If it takes T time units to perform this operation on each sample, then sequential processing the time to apply that operation to all samples, is N times T. If we have a cluster of four processors, we can distribute the data across the four processors. Each process operates on the dataset subset of N over four samples. The processing of the four subsets of the data is done in parallel. That is, the subsets are processed at the same time. The distributed approach's processing time is approximately N over four times T plus any overhead required to merge the subset results and maybe shuffle them (Figure 8.5). This is a speedup of nearly four times over the sequential approach [13].

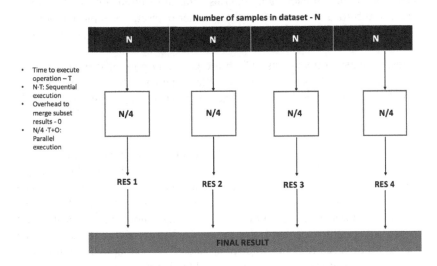

**Figure 8.5**  Speed up four times by dividing data into four subsets

A distributed computing platform such as Spark or Hadoop, scalable machine learning algorithms, uses the same scale-out approach. Data is distributed across different processors, which operate on the data subsets in parallel using map-reduce, and other distributed parallel transformations. This allows for machine learning techniques to be applied to large volumes of data. The Spark and its scalable machine learning library, MLF, provides a way to apply machine learning to big data. This is the processing of machine learning, where the data resides. Furthermore, it is called the Big Data Approach. However, it is also possible to scale up the machine learning algorithms by paralyzing such algorithms and using big data processing synchronically [14].

## 8.4    DATA TERMINOLOGY

### 8.4.1    DATA DEFINITIONS

Sample, variable, and feature are the keywords in machine learning with the following definitions:

> Sample or observation, which is usually a row of datasets, is an instance of recording an entity. For example, if the weather information is recorded hourly, each hour's information is considered an individual sample at each table's row.
>
> Variables are values in each sample, consisting of different information to represent a record. For instance, at the weather dataset, a sample includes variables such as sample ID and date, temperature, wind, and humidity at that hour.
>
> Features, attributes, dimensions, which are generally presented in columns of datasets, are the values to describe variables' characteristics. For example, wind in the weather dataset can be defined by its speed and direction.

### 8.4.2    DATA TYPE

The most typical data types are numeric and categorical data types, while others, such as time-series data and text, also exist in some literature. The most typical data types are numeric and categorical data types, while others, such as time-series data and text, also exist in some literature. Numerical or quantitative data are data points that take on exact numerical values. Numeric data that has a meaning as measurement can be categorized as continuous or discrete data. Continuous data can be any value within a range, while discrete data has distinct values. For example, the number of students, the number of transactions per hour, and the language are spoken discrete numerical data types; instead, the exam's score, house price, temperature, and heights are continuous data types. Categorical, qualitative, or nominal data types refer to non-numeric values that represent some quality or characteristics of an entity, for instance, city names, soccer player's positions, flower's names, and clothes colors. The categorical data does not have any explicit mathematical meaning; therefore, we cannot add them together or take the average. Samples in real-world data are usually multidimensional, meaning that each sample in the dataset is associated with different variables and dimensions with different data types.

### 8.4.3    DATASET IN MACHINE LEARNING

Machine learning models need to have three different datasets in order to perform well:

> *Training dataset:* after setting the hyperparameters, it adjusts the model's parameters so as to learn the input-output mapping; therefore, the most of the original dataset should be allocated in this part, usually 70–80% in average dataset size, if a problem is dealt with the very massive dataset this number

can be increased to 98%, for example when 10 million data is available for study.

*Validation dataset:* this dataset is used to see how well a trained model is working and determine if a trained model is overfitted or not. In simple words, it evaluates the generalization performance of a trained model and provides valuable information on how it is possible to increase model accuracy and performance. The validation dataset is usually 20–25% of the original dataset, while the original dataset is considered as a massive dataset it would be 1% of the original size.

*Test dataset:* this part of the dataset evaluates a model's performance on a new unseen data. It usually is between 5% and 10% of the original dataset. However, in a problem with a very massive primary dataset, it could be 1%.

It should be kept in mind that the test dataset should never be used in the training phase and used to tune the model. For example, in the cross-validation procedure to finish training. It should only be used on the trained model to make sure the model generalizes correctly.

## 8.5    MACHINE LEARNING MODEL

A machine learning model uses mathematical and statistical parameters and equations to discover the relationship between inputs and outputs. Figure 8.6 represents the model considering the input parameters, the equations that rework them, and the results (outputs).

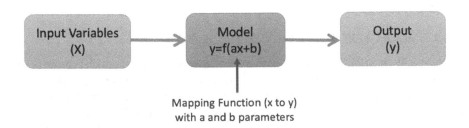

Figure 8.6    Machine learning performance

To put it another way, a machine learning model simply maps inputs to outputs by adjusting the mathematical parameters to improve the input-output mapping model's accuracy. In general, building the machine learning models, includes two phases:

Training phase: in this phase, the model adjusts the parameters considering the training data to minimize errors.

Testing phase: the new unseen data test the learned model [16].

The primary purpose of building a machine learning-based model is to have a model that works well on both training and test datasets as shown in Fig. 8.7 and 8.8 [17].

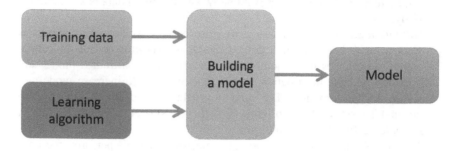

**Figure 8.7**   Training phase

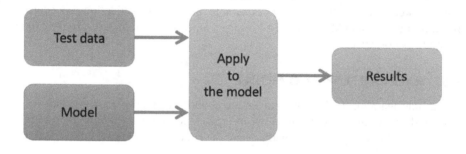

**Figure 8.8**   Testing phase

## 8.5.1   MODEL PERFORMANCE ANALYSIS (ERROR)

The errors in machine learning models are defined depending on which type of problem it tries to answer. Simply, the error is when predicted values or classes are different from the true ones. The measurement or criteria for calculating error in classification and regression algorithms are discussed in the following.

There most common metrics to calculate model performance for regression algorithms are mean square error (MSE), root mean square error (RMSE), and mean absolute percentage error (MAPE). In many cases, root mean square error is favorable to use since taking root of the error is proportional to the variable's units and gives us a better understanding of its range. There is also an R-squared error measurement to see how correlated the prediction values are with the actual values. It ranges between 0 and 1, and higher values mean the trained model has been fitted on data with better accuracy. In the classification algorithms, the precision of models defines as if predicted labels are the same as the actual ones there is no error. Otherwise, if the predicted class is different from true class, there is an error. In this case, the total error is calculated on how many classes are mispredicted over the entire dataset as a percentage.

Generally, the error in training data or error rate is referred to as training error, and the error on test data is called test error. The training and testing error not only on how well a model is working but also provides valuable information about why a model is not working well and what could be a source of a problem. If both of them are low and in the same range, it means a model's generalization, how well the model performs on the new data, is perfect and can be trusted on new data. Remember that any machine learning-based model's ultimate goal is to generalize well on new unseen data with the same probability distribution as what is trained. Sometimes test error is also called a generalization error since it indicates how well the model generalizes new data. The other situation relates to when a proposed model has very low training error and high generalization error, which is called overfitting. In this case, a model has learned to adjust parameters in such a way that governs the noise in the training set, rather than determines governors equations of the dataset [18]. Figure 8.9 shows an overfitted model, where the training data and input-output mapping are presented as points and as a cure, respectively. The model has learned the appropriate forms of data in the left figure, and the mapping curve follows a trend of data. On the other hand, the right figure illustrates the overfitting example in which the model has learned to model and follow the noise in samples. In this case, instead of finding the general trend of input-output pairs, the proposed model seeks to take every data point. It is good to mention that in real machine learning applications, while the model is training, the training and testing error are depicted together to follow the model's accuracy.

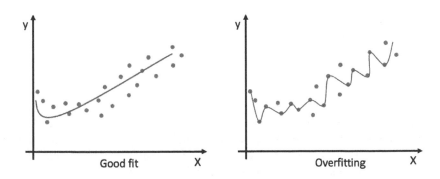

**Figure 8.9**  Modeling noise in the dataset (left plot is robust to noise)

The overfitted model does not generalize good enough to predict unseen data correctly; it only performs well on training data, and such a model is not useful and trustworthy for any real applications. The last situation happens when the model works poorly on both training and test datasets, called underfitting, where both training and test errors are high. Overfitting and underfitting are undesired situations that fail the machine learning-based model to be applicable in the real world. The desire solution is when both training and testing errors have low values, and the model can generalize well on the new datasets.

## 8.5.2 VALIDATION SET

A validation set is the part of the training dataset that is used to measure the accuracy of a model in the training phase. A validation dataset held back from training a model is different from the test set, as shown in Figure 8.10. However, in some literature, the validation dataset and testing dataset are considered the same, and the original dataset is not divided into three phases to build a machine learning model. One of the validation set advantages is to terminate training the model to avoid overfitting and improve generalization performance before testing the model with an unseen testing dataset.

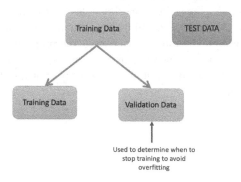

**Figure 8.10**   Validation set

As the model is training, both training and validation errors should be under observation. Figure 8.1 shows the validation error and training error while the number of nodes in the decision tree increases. If the number of nodes is continuously increased in the decision tree, the model's complexity also increases, and training error decreases, while a model probably becomes overfitted. It means that after adding a certain number of nodes, the model started losing generalization capability even if the validation error decreases [19].

One solution is to look at the training-validation error pairs and determine when to stop training before the model starts to be overfitted. For example, in the decision tree model, we can truncate the increment of the node's number, shown with a red line, exactly when the validation error starts to diverge the training error, as shown in Figure 8.2.

It is possible to create the validation set in several ways in order to avoid the overfitting problem. The most well-known ones are as following:

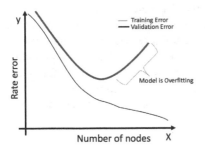

**Figure 8.11** Rate error vs. number of nodes

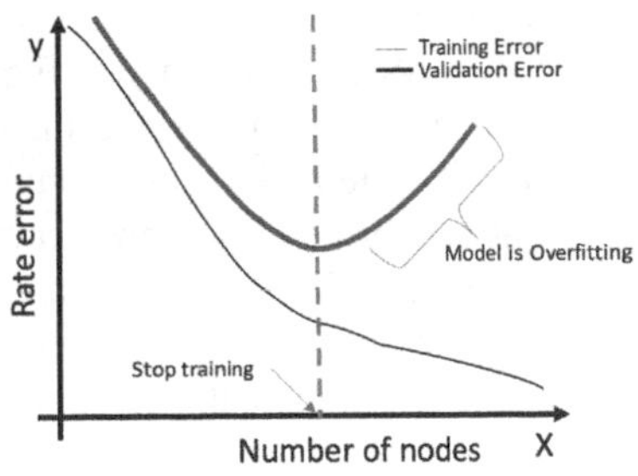

**Figure 8.12** Stop training threshold

*Holdout method:* in this method, which is mostly applied in machine learning tasks, some parts of the original dataset, is selected as validation dataset, as shown in Figure 8.13. The training and holdout set errors are calculated at each epoch during the training phase and plotted together at the end of the training phase. The training dataset should be more than data reserved for as a holdout validation set, usually 20% of the total dataset. Moreover, the training and holdout sets should have the same data distribution to avoid misleading results.

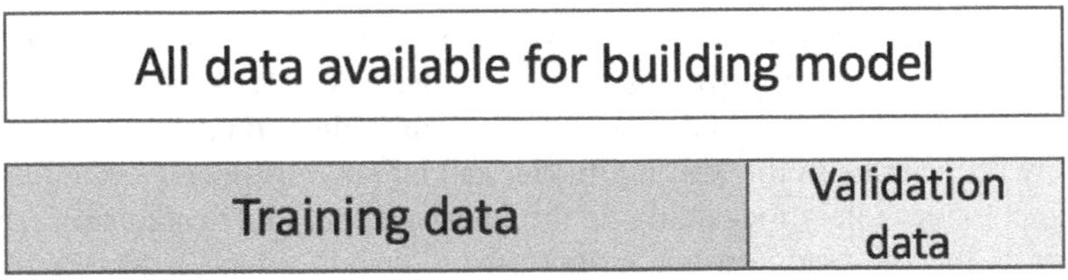

| All data available for building model | |
|---|---|
| **Training data** | **Validation data** |

**Figure 8.13**   Training and validation dataset in holdout method

*Random subsampling or repeated holdout method:* in this method, the holdout method is repeating many times by randomly picking samples from the original data to create training, validation, and test datasets. The reaping holdout method decreases the overfitting chance and increases overall generalization performance. This method's possible disadvantage is due to the possibility of selecting some samples more than others for training [20].

*K-Fold Cross-Validation:* this is a powerful method to avoid overfitting problems, especially in classification problems. The data is segmented to k number partitions, as shown in Figure 8.14. In each iteration, one segment is used as the validation set and the others as the training set. This is the most favorable method to select the best model in practice. K-fold cross-validation method provides a more structured approach to split the original dataset into training and validation datasets so as to avoid the overfitting problem by giving the variability in overall performance compared to suing an individual partition of the data.

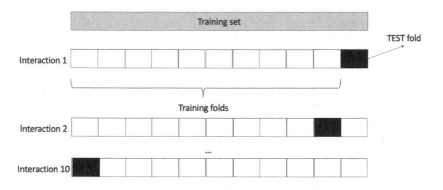

**Figure 8.14**   K-fold cross-validation method

*Leave-one-out cross-validation:* it is a particular case of cross-validation where the number of partitions (k) equals the dataset's size (N), number of instances, or samples in the dataset. Thus, the validation data set has precisely one sample in the learning model and all other samples as a training dataset for each iteration, similar to the statistical jack-knife estimation. The error on the validation set is not only used to decide when to stop training and avoid overfitting but also is used to determine the model's generalization performance.

### 8.5.3   MODEL PERFORMANCE'S EVALUATION METRICS

The metrics that are used in each machine learning algorithms are different from each other. In the following, the most used metrics in classification and regression algorithms are discussed.

### 8.5.3.1 Classification Metrics

The model's performance can be evaluated with different metrics. In classification algorithms, four various error types, reported in Table 8.1, are mainly used in evaluation metrics.

**Table 8.1**

**Different Types of Error**

| True Label | Predicted Label | Error Type |
|---|---|---|
| Yes | Yes | True Positive (TP) |
| No | No | True Negative (TN) |
| No | Yes | False Positive (FP) |
| Yes | No | False Negative (FN) |

The accuracy rate is broadly used as the evaluation metric by dividing the number of the correct prediction by the whole number of predictions [21]. In the following formulas for the accuracy and error rates, TP, TN, FN, and FP are the summation of predictions in each category.

$$\text{Accuracy rate} = \frac{\text{Number of correct predictions}}{\text{Total predictions}} = \frac{TP+TN}{TP+TN+FP+FN} \quad (8.1)$$

$$\text{Error rate} = \frac{\text{Number of incorrect predictions}}{\text{Total predictions}} = \frac{FN+FP}{TP+TN+FP+FN}$$
$$= 1 - \text{Accuracy rate}$$

Besides accuracy and error rate metrics, precision and recall metrics are also widely used when an imbalance class of prediction is concerned. The precision metric measures the exactness of prediction since it considers the percentage of predicted outputs as positive, as presented as follow:

$$\text{Precision} = \frac{TP}{TP+FP} = \frac{\text{Positive samples correctly predicted}}{\text{All samples predicted as Positive}} \quad (8.2)$$

On the other hand, the recall metric measures predictions' completeness since it considers the percentage of predicted positive outputs that a model identified correctly, compare to all true positive. Which is defined as follow:

$$\text{Recall} = \frac{TP}{TP+FN} = \frac{\text{Positive samples correctly predicted}}{\text{All samples with true label Positive}} \quad (8.3)$$

Additionally, the F-measure is another metric, which is the combination of both precision and recall metrics. There are different F-measures types depends on the weight of the precision and recall values on its formula. The most common ones are reported below:

*F1 measure:* both precision and recall values have the same weighted average, which is presented in the following equation:

$$F1 = 2 \times \frac{\text{Precision} \times \text{Recall}}{\text{Precision} + \text{Recall}} \qquad (8.4)$$

*F2* measure: recall's weight is higher than precision's weight
*F0.5* measure: precision's weight is higher than recall's weight

The value ranges from zero to one, with higher values giving better classification performance [22].

A confusion matrix is a table to summarize the misclassification error. It is used to represent a classifier's performance while the test dataset's actual values are known [23]. It is relatively simple to understand, however, its terminology sometimes can be complicated to implement. Based on actual values and predicted ones, four different possibilities are explaining as follows:

The true positive (TP): this cell includes samples that are correctly predicted as positive.
The true negative (TN): this cell includes samples that are correctly predicted as negative.
The false positive (FP): this cell includes samples that are incorrectly predicted as positive.
The false negative (FN): this cell includes samples that are incorrectly predicted as negative.

Each cell has the count, or percentage of samples, with each type of errors as reported in Table 8.2.

**Table 8.2**
**Confusion Matrix**

|  |  | Predicted Class Label | |
|---|---|---|---|
|  |  | Yes | No |
| True class label | Yes | True Positive (TP) | False Negative (FN) |
|  | No | False Positive (FP) | True Negative (TN) |

In the confusion matrix, the true positive and true negative values, inputs with correct outputs predictions, are presented in the diagonal table. On the other hand, the off-diagonal values relate to misclassification outputs, inputs with incorrect outputs predictions. Therefore, the higher summation of diagonal values or the smaller summation of off-diagonal values indicates the better the classifier learning model's overall performance. In the confusion matrix, the metrics are defined as follows:

*Accuracy rate:* the summation of the diagonal values divided by the total number of predictions.

*Error rate:* similarly, the summation of the off-diagonal values divided by the total number of predictions.

The confusion matrix provides very important information regarding a classifier's performance that it is easily understood the misclassifications' kinds that the model is making by looking at it. For example, when classifying the true negative or positive labels is problematic for the model. To sum up, the confusion matrix summarizes the various classification error types in a table, in which the diagonal and off-diagonal values are used to calculate different performance metrics and evaluate a classifier's overall performance of a learning model. Moreover, it represents which type of mis-classifications are the most problematic for a classification model.

### 8.5.3.2 Regression Metrics

The regression evaluation metrics are slightly different from the classification metrics because regression learnings output continuous values within a given range. The different metrics used to evaluate the regression's performance include Mean Squared Error (MSE), Root-Mean-Squared-Error (RMSE), Mean-Absolute-Error (MAE), Mean absolute percentage error (MAPE), and $R^2$ or Coefficient of Determination.

Mean Square Error (MSE) is widely used to evaluate regression learning results since it is differentiable and can be optimized easier. It is defined as the average of the squared difference between the actual values and predicted values, as presented in equation below, where n is total prediction number. Due to the squares of the differences, it considers even a small error and punishes a model for overestimating how bad a learning performance is.

$$MSE = \frac{1}{n}\sum(y-\hat{y})^2 \tag{8.5}$$

Root Mean Squared Error (RMSE) is the utmost broadly used metric to evaluate the regression's performance. It is the square root of the average of the squared difference between the actual values and predicted values, equation below. It is the most preferred metric since it highly penalizes large errors, mostly harming a model performance. Thus, RMSE is used when large errors are seriously taken into account and are undesired.

$$RMSE = \sqrt{\frac{1}{n}\sum(y-\hat{y})^2} \tag{8.6}$$

Mean Absolute Error (MAE) is the average of the absolute difference between the actual value and predicted one, equation below. This metric, which is robust to outlier and high error values, unlike the MSE, does not penalize errors extremely. Since MAE is a linear score, meaning that all difference has the same weight in the metric, it is not proper for applications when the outliers are critical.

$$MAE = \frac{1}{n}\sum|y-\hat{y}| \tag{8.7}$$

Mean absolute percentage error (MAPE) measures the accuracy as a percentage by calculating the average of the summation of all difference between the actual value and predicted value divided by the actual value. In other words, it is the average absolute percent error for each prediction. MAPE is commonly used for reporting the model's performance, while there are not many zero values and no extremes to data.

$$MAPE = \frac{1}{n} \sum \left| \frac{y - \hat{y}}{y} \right| \tag{8.8}$$

Coefficient of determination or R-squared ($R^2$) is the portion of the variance in the current model predictions performance from the constant baseline performance, presenting how much a model is better than the benchmark or other baseline model. The benchmark is chosen by taking the mean of the data and a line representing the mean. $R^2$ is a scale-free metric, thus even the values are extremely low or high, this metric always is less or equal to 1.

$$R^2 = 1 - \frac{\sum(y - \hat{y})}{\sum(y - \bar{y})^2} \tag{8.9}$$

## 8.5.4   MACHINE LEARNING ALGORITHMS

As it is mentioned before, machine learning algorithms can also be categorized into classification, regression, clustering, and association analysis. In this section, the most common supervised learning algorithms, classification, and regression are introduced.

### 8.5.4.1   Classification Algorithms

There are several classification algorithms; the most commonly used ones are presented as follows [24–26]:

**Logistic Regression:** Logistic regression is widely used for binary classification tasks where an output belongs to one class or another (0 or 1). In this algorithm, a threshold is defined to indicate examples will be labeled into which class using hypothesis and logistic function (usually sigmoid curve). The hypothesis determines the likelihood of events to generate data and fit them into the logarithm function that forms an S-shaped curve called sigmoid. Then, the logarithm function is used to predict the class of new inputs. Logistic regression is also applied to multiclass classification problems by applying the one versus all strategy, however, it has better performance on binary classification tasks.

**K-nearest neighbors (KNN):** This algorithm is one the most basic yet broadly used classifier, which without making any assumptions on data distribution, finds the data with similar characteristics and group them in the same class. The groups are constructed by considering the attributes of the neighboring samples. It is used in real-life problems in several applications such as data mining, pattern recognition, and invasion detection.

**Naive Bayes:** This technique is one of the most potent classification algorithm based on an extension of Bayes theorem, assuming each feature is independent to capture input-output relationships. The Baye's Theorem compares the probability of happening of an event to what has already happened, for example, the probability of having a fire (event A) while the weather is hot (event B, which is present) [27]. The naive algorithm is simple to implement and quick to predict labels of new inputs. Additionally, when a domain knowledge confirms the feature independence, with less data, it has a better performance than other classification algorithms such as logistic regression. On the other hand, in real life, it is not easy to have data with entirely independent features; moreover, when there is an input that was not follow up in the training phase, the algorithm assigns zero probability, and it does not classify this input in any group. Generally, this technique is used in various applications such as text classification, spam filtering and so on.

**Support Vector Machine (SVM):** This algorithm is widely used in classification tasks while also applied for regression problems. SVM's main idea is to transfer data to higher n-dimensional space to find and ideal hyperplane to differentiate classes. In simple words, these support vectors are coordinate of a new n-dimensional coordinate system. This method is common to use for binary classification, however, it is computationally expensive and slow in the big data domain.

**Decision Tree (DT):** This algorithm is based on different hierarchical steps that lead to certain decisions. It applies a treelike structure to represent repetitious decision paths with induction and pruning steps. In the induction step, the tree structure is built, while, in the pruning step, the complexities of the tree are reduced. The inputs are mapped to outputs by traversing each path through different branches of the tree. DT is the powerful classification tool and simple to structure with quick performance. However, with even small variations in data, DT can become unstable. Furthermore, it can easily become overfitted, especially in a thorny tree with many branches and conditions, thus, do not generalize well on new inputs. Regularization, bagging, and boosting techniques are usually used to avoid overfitting problem in DT.

**Random Forest (RF):** This classifier is very similar to the decision tree, which instead of having only one tree, it uses several decision trees. This technique can be applied in more massive data set in order to classify data or measure the importance of each feature in the final decision. In many applications, the random forest is preferred over the decision tree because it can be more accurate and overcome DT's overfitting problem. However, this technique is not easy to implement since it has a complex structure, and it is not recommended for real-time prediction purposes because, in general, it is the slower model.

There are many other classification algorithms, but these are fundamental algorithms that are commonly used and form other algorithms for classification.

### 8.5.4.2 Regression Algorithms

There are several regression algorithms (numerical or continuous value prediction) [28, 29]; the most commonly used ones are as follow:

**Linear Regression:** In simple words, this technique tries to find the fittest straight hyperplane to the data. It is commonly used when there are linear relationships between variables, and it can avoid overfitting by regularization techniques such as LASSO, Elastic-Net, and Ridge. However, it is not flexible in finding the best solution for non-linear relationships in variables and complex patterns.

**Regression Tree:** This technique has the same hierarchical structure as the decision tree; instead, it takes the numerical values as an input. The branching procedure not only maximizes the learning gain but also learns non-linear relationships between variables. Even if this method is robust to outliers and easy to implement, it is prone to overfitting problems. In addition to the regression tree, Random Forests (RF) and Gradient Boosted Trees (GBM), which are commonly used Ensemble methods, are also applied in numerical predictions and have better performance concerning overfitting problem.

**Deep neural network:** Deep neural network or multi-layer neural network is widely used in every domain since it can capture complex patterns, which can also be used as a classifier. The non-linear relationships between features are learned by non-linear activation functions and hidden layers between input and output. There are several techniques and methods to improve the neural network's performance as well as different advanced neural network-based models such as Convolutional Neural Network (CNN) or Recurrent Neural Network (RNN). The learning engine in self-driving cars, speech recognition, computer vision, natural language processing, and disease diagnosis is the deep neural network. However, unless other algorithms, the intense knowledge in how to tune a neural network model is needed to have a perfect neural network model. Besides, neural network models work well in the big data domain, and usually, they are very computationally expensive methods.

# REFERENCES

1. http://courses.csail.mit.edu/6.036/
2. Andrieu, C., N. de Freitas, A. Doucet, and M. Jordan (2003). An introduction to MCMC for machine learning. Machine Learning 50, 5–43.
3. https://www.sas.com/en_us/insights/analytics/machine-learning.html
4. https://www.forbes.com/sites/bernardmarr/2016/09/30/what-are-the-top-10-use-cases-for-machine-learning-and-ai/#570b7ea494c9
5. Bengio, Y. (2009). Learning deep architectures for AI. Foundations and Trends in Machine Learning 2(1), 1–127.
6. Alpaydin, E (2004). Introduction to Machine Learning. MIT Press.
7. http://www.monetate.com/resources/learn-about-personalization
8. Bar-Shalom, Y. and T. Fortmann (1988). Tracking and Data Association. Academic Press.
9. Hastie, T., R. Tibshirani, and J. Friedman (2009). The Elements of Statistical Learning: Data Mining Inference and Prediction. Springer.
10. Bishop, C. M (2006). Pattern Recognition and Machine Learning, Springer.
11. Murphy, K. P (2012) Machine Learning: A Probabilistic Perspective, MIT Press.
12. Bekkerman, R., M. Bilenko, and J. Langford (Eds.) (2011). Scaling Up Machine Learning. Cambridge.
13. Bakker, B. and T. Heskes (2003). Task clustering and gating for Bayesian multitask learning. Journal of Machine Learning Research 4, 83–99.
14. Argyriou, A., T. Evgeniou, and M. Pontil (2008). Convex multi-task feature learning. Machine Learning 73(3), 243–272.
15. https://spark.apache.org/docs/1.1.0/mllib-guide.html
16. Berkhin, P (2006). A survey of clustering datamining techniques. In J. Kogan, C. Nicholas, and M. Teboulle (Eds.), Grouping Multi-dimensional Data: Recent Advances in Clustering, pp. 25–71. Springer.
17. Ackley, D., G. Hinton, and T. Sejnowski (1985). A learning algorithm for Boltzmann machines. Cognitive Science 9, 147–169.
18. Brochu, E., M. Cora, and N. de Freitas (2009, November). A tutorial on Bayesian optimization of expensive cost functions, with application to active user modelling and hierarchical reinforcement learning. Technical Report TR-2009-23, Department of Computer Science, University of British Columbia.
19. Choi, M., V. Tan, A. Anandkumar, and A. Willsky (2011). Learning latent tree graphical models. Journal of Machine Learning Research vol. 12, pp. 1771–1812.
20. Ricky Ho. "Big data machine learning: patterns for predictive analytics." DZone Refcardz 158, no. 4 (2012).
21. Betz, M. A. and K. R. Gabriel (1978) "Type IV errors and analysis of simple effects". Journal of Educational Statistics 3, (2), 121–144.
22. Caruana, R. and A. Niculescu-Mizil (2006). An empirical comparison of supervised learning algorithms. In International Conference on Machine Learning, pp. 161–168. 2006

23. Bottou, L (2007). Learning with large datasets (nips tutorial).
24. Jindal, A., N. Kumar, and M. Singh (2016). A data analytical approach using support vector machine for demand response management in smart grid. In 2016 IEEE Power and Energy Society General Meeting (PESGM) (pp. 1-5). IEEE.
25. A. Jindal, A. Dua, K. Kaur, M. Singh, N. Kumar, and S. Mishra. (2016). Decision tree and SVM-based data analytics for theft detection in smart grid. IEEE Transactions on Industrial Informatics 12(3), 1005–1016.
26. Jindal, A (2018). Data Analytics of Smart Grid Environment for Efficient Management of Demand Response (Doctoral dissertation, Thapar Institute of Engineering and Technology Patiala).
27. Devroye, L., L. Györfi, and G. Lugosi (1996). A Probabilistic Theory of Pattern Recognition. New York: Springer.
28. Althobaiti, A., A. Jindal, and A. K. Marnerides (2020). SCADA-agnostic power modelling for distributed renewable energy sources. In 2020 IEEE 21st International Symposium on "A World of Wireless, Mobile and Multimedia Networks" (WoWMoM) (pp. 379-384). IEEE.
29. Jindal, A., G. S. Aujla, N. Kumar, R. Prodan, and M. S. Obaidat (2018). DRUMS: Demand response management in a smart city using deep learning and SVR. In 2018 IEEE Global Communications Conference (GLOBECOM) (pp. 1-6). IEEE.

# Section V

---

*Case Studies and Future Directions*

# 9 The Use of Machine Learning Techniques for Monitoring of Photovoltaic Panel Functionality

*Haba Cristian-Gyozo*
Faculty of Electrical Engineering Iasi, Romania

## CONTENTS

## 9.1 INTRODUCTION

Solar energy is one of renewable energy resources designed to replace more and more the energy produced with fossil fuels that have a negative impact on the environment. Photovoltaic systems, as one of the important alternatives for converting solar energy into electricity, are expected to provide solid support not only for a future sustainable development but also for the creation of new jobs, both at the place of production of these systems and at the sites where these systems are installed [5, 27].

The current trend of migration of energy production from the field of classical resources to the field of renewable resources is confirmed by the increase in the number of installed photovoltaic panels worldwide [14]. The desire to reach a sustainable energy production system stimulates a whole series of researchers, companies, local or national authorities and also organizations at the transnational level, to make their creative contribution to the implementation of new renewable green technologies.

There are two trends in the development of renewable resources [2, 18]. The first one involves the creation of large localized generation systems, usually outside cities. The second envision the installation of smaller power systems that can be distributed within the city, allowing the expansion of generation capacity through local generating systems. It must be kept in mind that technologies are changing at an accelerated pace, therefore, the selection of the right renewable generation technology to be used, that should stand a reasonable amount of time in order to obtain a return of investment, is not an easy task.

Regarding the photovoltaic technologies, one development direction involves adapting house building to photovoltaic technology to use the energy generated by them [21]. Another direction tends to integrate new technologies in the house parts in such a way that when classic parts are replaced (e.g. roofs, walls) the look of the house remains merely the same. This solution also has consequences for the methods and technologies used in monitoring the photovoltaic systems which also must be made ubiquitous.

From the point of view of energy generation conditions, production systems can be considered to fall into two categories: those in which the conditions are constant and those in which the conditions depend on the environment and, as a result, may undergo significant changes in longer or shorter intervals of time. Thermal power plants and hydropower plants are part of the first category, while photovoltaic panels and wind turbines are part of the second category.

Due to the operation in the external environment, the photovoltaic panels are exposed to weather changes, changes that in some places and in certain periods can have extreme manifestations with a direct effect on the integrity or proper functioning of their components. In order to ensure the recovery of investments and even the obtaining of some benefits, the photovoltaic systems are designed to face these changes of the working conditions over long periods of time, but the part of their monitoring and maintenance, plays an equally important role.

One of the important conditions that a photovoltaic panel must meet to ensure maximum efficiency is to capture the maximum amount of incident solar radiation to be converted into electricity inside photovoltaic cells. The fulfillment of this condition is influenced by several factors, among which we mention the positioning of the panel, the lack of obstacles in the path of radiation from sun to the PV cell, the radiation transmission coefficient from the incident surface through the protective layers to the semiconductor junctions where the conversion takes place.

The reduction in the amount of radiation received by photovoltaic cells can be caused by shading, by the obstruction of cells by materials (water, snow, dust, sand, mud) or objects of different sizes (vegetation in form of hay, leaves or tree branches, plastic films, other garbage) or due to the decrease in the transparency of the protective layers. These changes are directly reflected in the operation of the panels whose performance decreases, numerous studies being conducted to identify as accurately as possible these negative effects [3, 8, 26].

In order to remove negative effects as soon as possible, the defective operation status of the PV panels must be detected at an early stage and properly identified to initiate the appropriate maintenance actions.

The rapidity of identification depends on the location and size of the PV park, methods used (visual inspection which depends on visibility and weather conditions and experience of inspection person or other automated methods which can use system local or global information).

Correct prediction of the occurrence of a defective situation can help in taking preventive action in advance, limiting the decrease in PV panel efficiency and therefore of its output. Recently, machine learning techniques have gained attention and started to be used in different aspects related to PV panels.

In [39] the authors are presenting a research in which the Gaussian process is used in computing panels maximum power point (MPP) when there is a variation of the environment parameters. As the output of the PV panels depends on the amount of solar radiation, estimating the Daily Mean Solar power can be used to predict the resulting generated power.

Machine learning techniques can be very efficient also for classification purposes [4, 20, 22]. In the area of PV applications, these techniques have been used for classification of events occurring in networks that include PV systems [33], classification of defective operation of solar panels caused by breaks, dust, snow or weeds [6, 35]. If such events result in reduction of stability and controllability of the system, it is important to detect them as soon as possible and take rapid measures to correct the situation. Different models have been proposed and studied, the use of tensor flow [15] and artificial neural networks [19] resulted to have given the best results.

Machine learning techniques can be applied if there is a lot of data that can be used to train and validate a model that can then be used for prediction [28]. In the case of PV systems, the advance of technology allows on one side the embedding or addition of relevant sensors and on the other side, the provision of data from the conversion and controlling devices of the system. In addition, as environment plays an important role in PV system operation, availability of environmental data increases the options in combining different data sources in order to build and train better models. In [17] it is proposed a development system that aims at providing the functionality of integrating different information flows in order to be used for monitoring the PV system operation and early detection of changes in system integrity.

In this book chapter a review of the latest research done in the area of applying ML techniques for identification of possible degradation of photovoltaic panel operation will be presented at the beginning of the chapter. The main causes of panel operation degradation will be then presented and the problems that may appear when this situation occurs in a PV park. Next, a monitoring system architecture will be presented based on authors recent research that involve different sources of information. These include sensors which are part of PV panel system, sensors distributed in the PV park, meteorological stations in the proximity of the park, other sources of information available on the internet (weather channels, environment monitoring feeds as well as historical data). The results of applying different ML techniques will be presented and conclusions about the efficiency of applying these techniques in the frame of the future Internet of Energy will be drawn.

## 9.2  SOLAR PANEL MONITORING

Today PV parks are complex systems based on advanced technologies that support not only solar energy conversion, interaction with load, storage systems and network grid but also generation of useful information which is communicated via communication channels to be used for monitoring, sustain of their daily system operation and interaction with the working environment and human personnel.

As shown in Figure 9.1 the PV panel monitoring systems monitor the PV panels in their environment, perform continuous assessment of their physical and functional integrity (remote monitoring) and gather data from different sources in order to analyze system's behavior and to improve operational performance.

The aim of PV monitoring system is to improve system operational conditions and functionality, as well as to ensure system safe and secure interaction with other systems. Beside, in the case of PV parks it is useful to continuous monitor the large number of PV panels as it constitutes a real support also for maintenance teams.

The devices that are generating data used for processing by the monitoring system can be part of the PV system or can be distributed in the local or enlarged environment (defining thus an intelligence at an environmental level). In this respect, a minimal list of characteristics that the monitoring system must implement includes compatibility with PV control systems, local and wide area connectivity, accuracy of collected data and security, all these in a frame of low power consumption.

The many possibilities of monitoring systems include:

Normal operation recognition (availability of sun radiation, identifying the good tilting angle, possibility of either consuming or storing the generated energy).

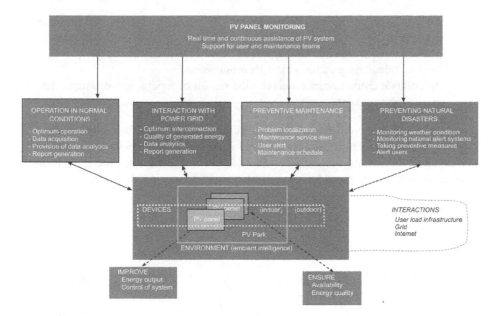

**Figure 9.1**    General view of PV panel monitoring

PV panel localization and identification (this includes physical identification of one or many PV panels and also their identification in databases, all these based on different available information). The accuracy of localization depends on the monitoring system complexity and can include different technologies such as visual identification, video monitoring, signature identification, machine learning techniques.

Event detection, recognition, classification and localization. Events can be of different nature (electrical, mechanical, and optical) therefore, the monitoring system should include a set of techniques associated to each class of events. This will provide information to build local or general situation awareness;

Behavior analysis, will include detection and identification of behavioral patterns based on a statistical predictive algorithm using different activity models (sunny-cloudy, day-night, summer-winter, load-no load etc).

Interaction with user and other systems (frequent use cases implying human-system or system-system interaction). Keeping these interaction at an appropriate level, allows the system to be aware of the local and overall context and have the capacity to make predictions regarding the situation evolution.

Besides their basically goal for increasing and monitoring the quality of system operation, the monitoring system may be useful also:

to assess the general status of the PV system (long term or continuous status monitoring);

to detect the emergency situations and subsequently elaborate appropriate suggestions for intervention (detection of damaged parts or entire subsystem, detection of hazards that can put the system or personnel in danger);
   to detect degrading status and/or its progression;
   to evaluate a maintenance intervention (result of repairs, substitutions, upgrades, improvements etc).

In Figure 9.2 is presented a general view of a PV panel monitoring environment identifying different aspects involved in the process, namely users, local and enlarged environments, technologies and devices that constitutes the ground basis for developing applications for implementation of monitoring solutions.

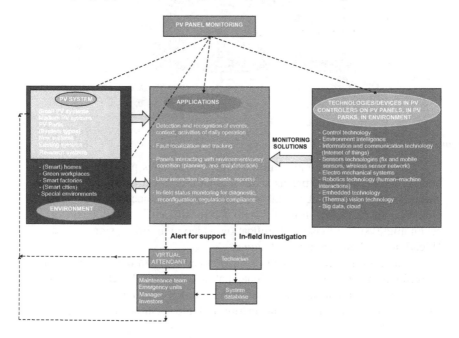

**Figure 9.2**   Elements and applications involved in PV panel monitoring

## 9.3   PHOTOVOLTAIC OPERATION DEGRADATION

The ability of PV panels to provide same power levels over their lifetime is important because it directly influence the capacity to supply a certain power needed by the load and to ensure the return of investment. Numerous studies have been carried out identifying the degradation rates both by doing indoor testing labs or in the outdoor field [25].

The reduction in operation or efficiency of the solar panels due to covering the solar cells with different materials is determined by the shadowing effect [37]. Figure 9.3 presents the cases where solar cells of PV panels are covered to a certain degree with sand or mud, resulting in the modification of their generating current and further on of PV panel operation. Figure 9.3 shows that cells can be covered at different degrees thus, some of them can bias the bypass cell, resulting in a reduction in the generated power. Depending of the PV panel internal architecture, bypass diodes can be provided for every cell or for a group of cells. In the second case, the shadowing effect of a single cell has a greater impact on the overall PV panel operation (an entire group of cells are bypassed) than in the first case [38].

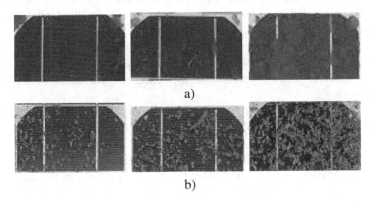

a)

b)

**Figure 9.3**  PV cells covered at certain degrees level with sand (a) and mud (b)

This interposition can be temporarily or can be of long term like the deposition of dust to the surface of PV cells [10, 13]. Several studies show the negative effect of dust deposition and that composition, and therefore color, can impact differently the PV cell degradation [26].

The reduction of PV panels' efficiency can be produced also by the effect of the air, water, salt or other impurities specific to regions located near the sea or oceans [40]. The main effects of this exposure is on the optical characteristics of solar panels because, the shielding glass can change color or even have its structure damaged after a certain period of time.

Some research has led to the development of additional devices that are attached to PV panels to remove impurities that can be deposited on the surface of the panels. These devices are based on various principles, the most common being those in which impurities are removed either with the help of air jets [41] or with the help of water jets. Other solutions consider the use of wiping or brushing systems, while another solution presented in [32] uses an electro dynamic screen. The cleaning systems can be individual, each panel having its own cleaning system or collective, in which case a single system is used, usually mobile, which performs the cleaning of several panels in a cleaning operation.

Another category of phenomena that can have medium to high impact on PV panel operation and also on their integrity are the meteorological phenomena. Among the most common phenomena of this kind we mention:

    hail storms;
    dust storms;
    ice storms;
    strong winds;
    tornados;
    storms with electrical discharges.

Heavy rain with hail can result in glass broken modules which evidently have important loss in power and yield. The problem is that modules that seem to look good, often have small damages most of the time invisible to the human eye, which cause them also significant loss in power.

PV panels and parks that face the power of tornados are usually severely damaged at panel, panel support and even PV park infrastructure level. Because these are evidently highly impacting phenomena, inspection and evaluation of the damages are mandatory.

Lightning strikes, whether direct or indirect, can result in both down time and expenses in identifying and replacing damaged parts [7].

Direct strikes, (rare), would destroy (melt) panels, inverters, etc.

If panels are not protected, lightning can produce damages whether it strikes directly the PV panels or in their nearby [7]. Such an event can produce damages both in the electrical circuit of PV system or in its structure. Any of this kind of damage will affect PV system operation generating loses due to down time needed to identify and fix the problem and replacement costs of damaged parts or entire PV panel.

## 9.4   PREVENTING MEASURES

In order to prevent or limit the effect of extreme events, it is important to obtain useful information about them (possibility of occurrence, moment of occurrence or impact, intensity, other characteristics) before their occurrence. This information may come from the park's own monitoring system or from other available sources of information.

Having such information available, various proactive measures can be taken to ensure the proper functioning of the system and the prevention of undesirable effects such as energy theft [23, 24].

## 9.5   REAL-TIME DATA ACQUISITION AND ANALYTICS

### 9.5.1   DATA SOURCES

Machine learning makes no exception when it comes to photovoltaic systems. In order to obtain the best possible models, a large volume of data is needed to be used for training the model and to validate it. With respect to PV systems, taking into account its highly interaction with the environment, there are numerous sources that can be used. In our study we have identified different categories of data sources as we can see below:

data received from sensors placed within the PV park. They can be mounted on the PV panels providing information on solar radiation, tilt angle, temperature, humidity, shocks, or they can be placed on other components of the photovoltaic park;

data received from PV panels and their control devices such as inverters, storage battery packs, power distributors, power lines, grid connectors etc;

data collected from small weather stations located in the PV park or in the region where the park is located;

data published by meteorological services that can be accessed from internet using dedicated application programming interfaces (APIs);

data from services computing or collecting astronomical data (in the case of PV parks we can be interested in data related to sun position, sunrise, sunset). More sophisticated studies can make use of data regarding cloud movement, storm or other extreme weather predictions;

data from cloud services where historical data are stored, integrated, processed and where real-time data can be processed, displayed and stored for future use;

data from geopositioning services that are useful for locating a specific PV park, PV panel or data sources (weather stations), for mapping them or for annotating data with geographic information;

data from alert systems provide information not only about severe weather phenomena but also about events that can affect PV park operation or can put them in danger.

### 9.5.1.1  Local Data Acquisition Systems

Local data acquisition system are relying on a set of sensors which are either embedded or mounted on the PV panel with the purpose of acquiring real-time local data to be used in controlling algorithms or monitoring system status. They are sometimes called weather stations because they measure temperature, pressure, dew point, wind direction and strength, level of illumination etc. They work in conjunction with gateways which allow them to connect to Internet and send acquired data to data centers or cloud services where data is analyzed, stored, process and displayed in order to support decision elaboration.

Usually controlled by a microcontroller, these systems can be extended to also collect information related to panel orientation with respect to solar radiation (the case of PV panels which include tracking position systems), can perform detection of a storm in the near region of the station providing also an estimation of its distance, speed, direction of movement and its evolving strength.

### 9.5.1.2  Meteorological Mini Stations

With people's increasing awareness of the importance of preserving the environment and the effects that human activity has on it, through various programs or individual initiatives, mini-stations have been installed in different geographical locations. Many of them are connected in various networks that manage them and that offer, sometimes free of charge and using a standard interface, the data collected in real time or historical data sets of different parameters.

One of these networks is the Weather Underground global community network which brings together data about weather or about air quality in the region where they are located. The network is based on individual initiatives all over the world, therefore, there is no uniform geographical distribution of stations. Figure 9.4 shows two regions in Romania, close to two cities of comparable size and population but with different number of mini-stations in the vicinity.

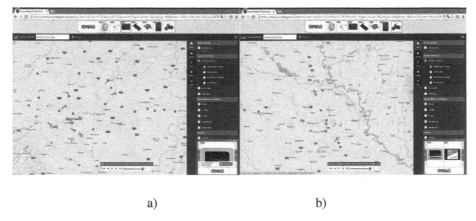

a)                                                    b)

**Figure 9.4**  Number of weather stations in two different regions of Romania a) Iasi (5 stations) b) Cluj-Napoca (15 stations)

There is a standard set of parameters measured by mini-stations including temperature, humidity, Dew point, precipitation level and pressure. Other parameters (like solar radiation) are measured only by some stations and published by their webpage or through corresponding API.

If such stations are available in the vicinity of the PV park, they can be used as data sources to feed the training, validation and prediction process, part of the PV monitoring system to be discussed in the next section.

Table 9.1, extended from [17] with two new operating stations in the region of the use case, provides a summary of data available from these stations and their characteristics.

### 9.5.1.3  Astronomical Data

Because the amount of power generated by a solar panel depends directly on the incidence of solar radiation, it is important to know the sun position. This is useful

**Table 9.1**

**Characteristics and Data Types Provided by Weather Stations Located in the Region of PV Platform**

| Weather Stations Characteristics | IIAI19 | IIAI20 | IIAI22 | IIAI40 | IISVLADI2 |
|---|---|---|---|---|---|
| Measurement of Temperature, Humidity, Dew Point, Precipitation, Pressure | Yes | Yes | Yes | Yes | Yes |
| Measurement of Wind Direction, Speed and Gust | Yes | Yes | Yes | Yes | Yes |
| Measurement of Solar Radiation | No | No | No | Yes | Yes |
| Measurement period (min) | 5 | 15 | 15 | 5 | 5 |
| Distance from PV platform (km) | 1.64 | 2.82 | 0.53 | 1.31 | 4.55 |
| Weather History in Graphs and Tables | Yes | Yes | Yes | Yes | Yes |
| Webcam | No | No | No | No | No |
| Elevation (m) | 39 | 104 | 30 | 45 | 99 |
| Position | 47.149° N, 27.572° E | 47.175° N, 27.574° E | 47.156° N, 27.6° E | 47.144° N, 27.582° E | 47.124° N, 27.633° E |
| Hardware | La Crosse Technology | Davis Vantage View | RaspberryPi | Other | Other |
| Software | Cumulus | weatherlink.com | weatherlink.com | EasyWeather | EasyWeather |

either to calculate or estimate the solar radiation received by PV panel during the daylight or, for panels with sun tracking system, to compute the tilt angle of the panel in order to receive the highest level of radiation possible. Studies show that even for fixed PV panels, the selection of optimum tilt angle for different periods of the year can result in important increase in energy output [30].

There are different available sun position calculators available on Internet, the most used being SunCalc.org (www.suncalc.org), NOAA Solar Position Calculator (https://www.esrl.noaa.gov/gmd/grad/solcalc), SunCalc.net (www.suncalc.net), SunEarthTools (www.sunearthtools.com). Besides knowing the sun position, it is important also to know the sunrise and sunset. With this information we can select the intervals of time where data are relevant to solar power conversion or when energy used in a PV based system is taken from storage batteries of from the grid.

Figure 9.5 shows data displayed at the location of PV park (selected for the use case) situated in the proximity of the Faculty of Electrical Engineering. The application programming interface (API) of the site can be used to obtain data at any location in the world at any date and hour by accessing a link with the following structure:

https://suncalc.org/#/lat,lon,zoom/date/time/objectlevel/maptype

where lat = latitude, lon = longitude, zoom = map zoom, date = date, time = time, objectlevel = object height for shadow calculation, maptype = type of map (0 = OpenStreet Map, 1 = Esri-Satellite, 2 = Esri-Topo, 3 = Esri-Street).

**Figure 9.5** Sun position calculator (www.suncalc.org) providing data for PV park near the Faculty of Electrical Engineering Iasi

#### 9.5.1.4    Cloud Services

Possessing accurate weather information at the right time has become a necessity for many businesses that get impacted more frequently by the recent climate changes and increase in the extreme meteorological phenomena [1]. Energy generation is one of the sectors that can be affected by weather events and renewable systems are among the most sensitive part of the energy segment.

The increasing amount of collected meteorological data and better accuracy of weather forecasts based on this data has increased users confidence in using this kind of data for analysis and making predictions important for different activities.

In the case of monitoring PV parks operation, meteorological data can be used to prevent damages of PV panels due to extreme phenomena, identification of long periods of non friendly conditions of operation, identification of disturbing transitory conditions, all that can result in a degrading capacity of PV panels operation.

Meteorological data are available now via different services, many of them accessible via Internet. From the existent sources, we have considered the ones that are freely available and those which provide an easy access to the data via a dedicated API.

Among the services we have analyzed we can mention AccuWeather, OpenWeatherMap (openweathermap.org), DarkSky (darksky.net), Troposphere (troposphere.io). The free versions of data access are limited to a certain number of data per day, the frequency of the data and the length of historical data that can be retrieved from their databases.

#### 9.5.1.5    Alerting Systems

As presented in section 1, PV can be damaged by extreme weather events. Therefore, in their monitoring process it is useful to take into account this type of events that may take place in the area where these panels are located, to be predicted, to be detected and to take appropriate measures.

These kinds of events are requiring special attention not only because they can cause material damages to the PV panels, but in some cases they can result also in loss of human lives.

Alerting systems have been developed by many countries and they provide useful information to administration bodies at different territorial levels, to state or private companies and even to individual persons. In recent years, there have been put into action programs such as the Severe Weather Information Centre of World Meteorological Organization, World Meteorological Organization's World Weather & Climate Extremes Archive from Arizona State University and others that integrate national or regional systems at a continental or even a global level. These systems allow the identification of extreme phenomena and their global alertness, because their effects can span (directly or indirectly) over large geographical areas, exceeding the boarders of a single country.

European National Meteorological Services have grouped together in the EUMETNET network which aims to provide a functional structure that can organize in

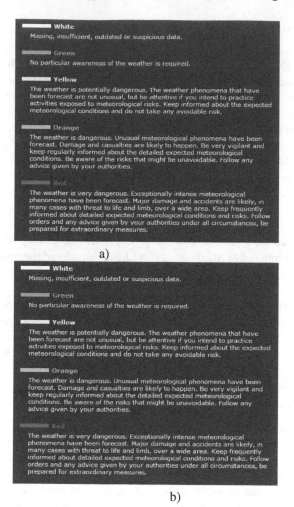

**Figure 9.6** Awareness types a) and levels b)

a co-operative approach EU Member states' activities in the meteorological area. This framework includes data gathering and processing systems, forecasts, reports, together with activities oriented to research, system development and training programs. The meteorological information issued by EUMETNET is made available to the European user through the dedicated website Meteoalarm.eu.

We focused on the Meteoalarm system because of the location of our study case. Other alert systems provide similar product and services via their associated websites. The Meteoalarm system can alert for possible occurrence of severe weather.

Figure 9.6 shows the available awareness types and levels with their associated icons and colors that can be displayed in maps associated to region of interest. The use of these symbols can quickly provide an overview of the existing weather situation in that region, highlighting on the map the points where extreme events occur or may occur.

From this set, we will look for the occurrence events that can impact PV panel operation such as severe thunderstorms, gale-force winds, heavy rain with risk of hail, heat waves, fog, snow or extreme cold conditions.

Figure 9.7 shows an alert for atmospheric instability that will result in heavy rain with hail and thunder storms in the region of Iasi County.

This information can be integrated with other information collected from weather stations which can include also storm detectors and meteorological services. The overall information can then be used for taking immediate decisions regarding preventive actions to avoid or reduce damages of the PV park or as a notification that at the end of the event, an inspection or an evaluation of the PV park status will be mandatory.

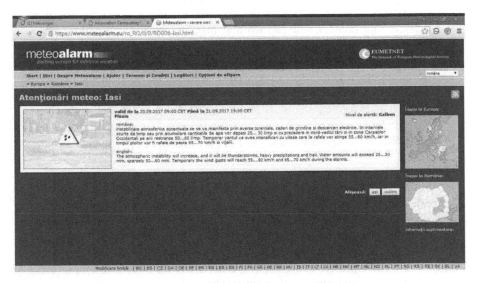

**Figure 9.7**   Alert of heavy rain with hail for Iasi County

## 9.6   MACHINE LEARNING TECHNIQUES IN PV PANEL OPERATION MONITORING

The increase of data available from energy source generation systems allows us to use machine learning techniques in order to find hidden although natural patterns in data that can provide a better insight of systems and processes helping one in making better decisions and predictions.

The use of machine learning algorithms is based on availability of data from which, by using different computational methods we can learn relevant information about a process directly from data. There is no need to have a mathematical model or a set of equations from which to extract the information. The model is obtained by learning which subsequently is used for making predictions for newly incoming data.

Based on different learning styles, machine learning algorithms can be divided in four different categories:

Supervised learning: this method is based on a training process for obtaining the model. The training process uses data sets including both correct and wrong predictions. Corrections are applied to the model when those predictions are wrong. The training is an iteration process that ends when the model, using the training data, reaches an expected level of accuracy.

Unsupervised learning: in this method datasets are not labeled, therefore, the model is obtained by first identifying existing patterns or structures in the input data. This can be obtained by reducing redundancy, by organizing data by similarity and extracting non visible rules.

Semi-Supervised Learning: in this method models are obtained based on datasets that contain both labeled and non labeled data. Therefore, algorithms must correct the model based on good and wrong predictions but also must identify structures in the labeled and non labeled data.

Reinforcement Learning: in this type of learning the machine is trained to make specific decisions. The training is based on a trial and error process in which data is feed continuously to the model. The machine is learning from the past datasets (experience) and adds knowledge that is used in making the best predictions in order to take the next best decisions.

Another classification of machine learning techniques can be obtained based on the method used for solving the problem.

A summary of machine learning techniques applied in PV system application found in literature is given in Table 9.2.

Applying machine learning technologies can bring a lot of benefits if the techniques are well implemented. There are some steps that one can take in order to develop a correctly machine learning approach:

1.  Gathering data. This is the step of acquiring data or identifying already

**Table 9.2**

**Application of Machine Learning in PV Systems**

| ML category | Method | Characteristics | PV Application |
|---|---|---|---|
| Classification | Logistic regression | Fast training Linear model Classes expressed in probabilities | Snow coverage [17] |
| | Support vector machines (SVM) | Robust Handles non-linearity | PV system performance [42], Solar radiation [43], Daily mean solar power [19], Fault Detection and Diagnosis [15], Solar energy prediction [44] |
| | Random Forrest (RF) | Fast training Handles categorical features Does not require feature scaling Captures non-linearity and feature interaction facilitate the reduction in the over-fitting | PV power generation [45] |
| | Naive Bayes | Good for text classification Assumes independent variables | PV system performance [42], Fault Detection and Diagnosis [46], Classification of Defective Solar PV Modules [35] |
| | k-Nearest Neighbors (k-NN) | Simple implementation. Are robust with regard to the search space | Fault Detection and Diagnosis [15], PV power generation [45] |
| | Decision table | Work on any type of data distribution, good for classification | Fault Detection and Diagnosis [15] |
| Regression | Linear regression | Prediction of continuous-valued output | Daily mean solar power [19] |
| | Regression tree | Partitioning of the input space | PV system performance [42] |
| | Gaussian regression | Prediction is Gaussian, | Maxim power point determination [39] |
| | Qunatile regression | Extension of linear regression, estimates the conditional median | Solar Irradiance Forecasting [47] |
| | Support Vector Regression | Based on similar algorithm as SVM | PV power generation [45] |

**Table 9.2**
**(Continued)**

| ML category | Method | Characteristics | PV Application |
|---|---|---|---|
| Clustering | K-Means | Automatic grouping of similar objects into sets (clusters) | Activity-based load disaggregation [33] |
| | Gaussian mixture models | A soft clustering method | Activity-based load disaggregation [33] |
| Artificial Neural Network Algorithms | Stochastic Gradient Descent Algorithm | Are efficient, easy to implement | Establishing series and parallel resistors for the one- diode model of a PV panel [48] |
| | Back Propagation Neural Network | Fine-tunes the weights of a neural net based on the error rate | Daily mean solar power [19] |
| | Other ANN | Variation of the ANN algorithms | PV power generation [45] |
| Deep Learning Algorithms | Convolutional Neural Network (CNN) | Learns and identify patterns from data Tuneable to obtain better and accurate results | Failure diagnosis [6], Short-term Solar Panel Output Prediction [49], Differentiate grid (including PV generation) events [50], Forecast of Solar Energy Production [50] |
| | Recurrent Neural Networks (RNNs) | Perform well in time series predictions as it remembers information through time | Home energy management [51] |
| Ensemble Algorithms | Boosting | Composition of multiple weaker models, component models are independently trained, whose predictions are combined into an overall prediction | Solar energy prediction [44] |
| | Weighted Average (Blending) | Composition is made based on weighted average | PV power generation [45] |
| Evolutionary algorithms | Multi gene genetic programming | Used in the search part of ML algorithms, based on stochastic search | PV system performance [42] |
| | Other GA | Blind search over the solution space, effective for some class of problems | PV power generation [45] |

available historical data. The quantity and quality of data are equally important for obtaining an accurate model.

2. Data preparation. Data can come from different sources with different characteristics (frequency, accuracy, range etc). Data must be put together and sometimes must be anonymized. Another important part of data preparation related to its split in two parts, the training set and the evaluation set. A typical split of the data is the one where training set is 80% and the evaluation set is 20% of the data set.

3. Choosing a model. Machine learning is an old but also hot topic; therefore, many models have been created to suit for different types of data such as text, images, sounds, sequences, numerical data and so on. It is important to choose the right model for available data. Sometimes is good to test different models and some techniques are based on combining several models.

4. Training. It is the step where the model is trained based on the training set. A better training increases the ability of the model to predict the result. A better training can mean using a larger training set but also using a more accurate labeling of the training set.

5. Evaluation. It is the step when the trained model is evaluated by testing it against the evaluation set.

6. Parameter tuning. Models have parameters that can be adjusted resulting in an improvement of the performance. Sometimes even small parameter tuning can result in a significant increase of performance.

7. Prediction. The last step is the one when the developed model is put to work and used to make predictions.

## 9.7 CASE STUDY OF SYSTEM FOR PV PANEL MONITORING

### 9.7.1 PHOTOVOLTAIC SYSTEMS IN ROMANIA

In Romania there have been started many solar park projects since 2005. The installation of photovoltaic parks has seen an irregular development with, for example, year 2019 reporting zero MW capacity installation. Even so, Romania is positioned slightly above the middle of the EU countries ranking in terms of installed solar photovoltaic capacity and also when considering electricity production from solar photovoltaic systems. The last Photovoltaic Barometer published by EUROBSERV'ER in April 2020 [14] shows that Romania has a cumulated solar capacity of 1385.8 MW. Given that Romania is part of the U.E. each kW of green energy is paid for by the government with green certificates, an additional reason not to miss this opportunity. In addition, the return on investment is estimated at 4–5 years, depending on the area, and the guarantee of solar panels (not taking into account as a calculation the rest of the equipment) is 20 years and increases with the improvement of the manufacturing process and the materials used.

The existent solar parks differ not only is sizes and generation capacities but also in types of solar panels installed and type of usage of produced energy. The average area of the solar parks is between 20 and 50 ha. Actually around 5000 ha of agricultural land are occupied with photovoltaic parks. Many of the parks were created by farmers who have found that solar energy can provide a certain energy independence and can be a solution, where the national system has not penetrated (in the mountains, in sheepfolds or in the fields), to produce energy for their own consumption or using solar energy to pump water for the irrigation systems.

Other farmers solved their electricity supply problems by developing a parallel business from the sale of surplus produced energy, something that farmers in Western Europe have been practicing for a long time.

Large area solar parks can be a solution for the cities if there are available areas near the city limits, something which is now difficult as the real estate are pushing hard these limits and tend to take any available space in order to develop large residential areas.

A solution for this situation is installing smaller photovoltaic parks in the city, the best solution is using the roofs of production units in the industrial area of the city. Some installations have already been put in place such as the one of a screw factory in Bacau, Romania that hosts the largest photovoltaic park in Romania, built on the roof. With an installed capacity of one MW, the plant will provide up to half of the company's energy consumption, following an investment of over one million euros.

The recently adopted legislative changes will encourage in Romania the construction of photovoltaic parks on the roofs of buildings, because such projects will no longer be able to be developed on agricultural land.

## 9.7.2   DESCRIPTION OF THE PHOTOVOLTAIC SYSTEM

The research aiming to develop the presented monitoring method was developed on the platform created with the occasion of project ENERED whose main objective was the development of the interdisciplinary research-development platform for efficient and sustainable energy (ENERED), within the "Gheorghe Asachi" Technical University of Iasi. The main result of the project was the implementation of a platform for five faculties of the university which enabled the carrying out collaborative research for quickly leading to results applicable in the field of energy and energy efficiency in the context of a sustainable development.

Part of this platform is the photovoltaic park built in the proximity of the Faculty of Electrical Engineering. A description of the park and its equipment is given in [16]. In few words, the photovoltaic park includes 9 PV panels arranged in three rows, with panels in each row differing only in the type of tracking system used.

Panels used for the PV park are of SS-BP250 type with 60 cells, glass laminated and embedded into an ethylene-vinyl-acetate membrane with an aluminum anodized frame.

Panels in one row are fixed (no tracking system) while the panels in the other row use one axis tracking systems form Deger Energie [12] and two axis tracking systems from the same manufacturer [11].

A summary of the panels tracking and control systems characteristics are summarized in Table 9.3.

**Table 9.3**
**PV Panels Control and Tracking System Characteristics**

| Panel Row | No. of PV Panels | No. of Inverters | Conversion and Control System No. of display and Inverter Configuration System | Tracking Type | Tracking system Elevation Angle | Elevation Drive (mm) | Rotation Angle | Power Consumption (W) |
|---|---|---|---|---|---|---|---|---|
| First | 3 | 1[a] | | Fixed | 15°/30° | - | - | - |
| Second | 3 | 1[a] | 1[b] | Single axis | 15°/30° | 850 | -45°/45° | 12 |
| Third | 3 | 3[a] | | Dual axis | 20°/90° | 1000 | 300° | 16 |

[a] OutBack Power inverters and charge controllers.
[b] MATE device (Outback Power and Technologies 2011).

**Figure 9.8** Prototype of a weather station including localization and lightning activity detection

When the park is not connected to the power grid, the energy is stored in a bank of 14 12V- batteries.

### 9.7.3   WEATHER STATION PROTOTYPE

Though some PV systems can include different sensors that collect data which are used in the control algorithms of sun tracking systems and conversion devices, these are not available to the user of the system. Therefore, in order to collect data local to PV panels, to be used in monitoring PV park operation, a weather station (WS) was developed. The WS which is described in [9] was developed based on a 32-bit microcontroller development board and includes sensors for temperature, humidity, thermal radiation, pressure, storm lightning activity and modules for wireless communication and GPS localization.

Data collected from different sensor can be displayed locally using an OLED display and can also be stored on an SD memory. The software for the station prototype was developed using the Arduino development environment. Figure 9.8 shows a picture of the WS prototype.

### 9.7.4   DATA SOURCES

#### 9.7.4.1   Data from PV System

Data from PV system are collected using the MATE device (part of the PV panels control system) that can provide real-time data about system status, battery status, and configuration data for the managed inverters and charge controllers. Data can be visualized on the MATE device, some data recorded can also be displayed on the device LCD panel. Through a system management HUB data for each inverter and charged controller can be also visualized on a web page accessible with any web browser. The access nevertheless is only local in the network of the faculty.

Additional information about data that can be accessed via the MATE device are presented in [16].

In order to collect and record data from the PV system and use it in the ML algorithms, an application was developed to run on a local server running Linux RedHat OS. The application is using *phantomjs* which is a scriptable headless web browser. The application consists of several scripts written in bash, calling *phantomjs* with an appropriate JavaScript script to access the page for every inverter and charge controller. The cost of the collecting system has been reduced by implementing the application on a Raspberry Pi single board computer running the Raspbian OS. Data is collected every minute and saved in files in csv (Comma Separated Values) format. Every file stores data for approximate one day (1440 reads).

### 9.7.4.2    Weather Ministations

There are 5 weather ministations as part of the Wunderground network in the nearby region of the PV platform (https://www.wunderground.com/weather/ro/iasi). Characteristics of these stations are given in Table 9.1. Figure 9.9 shows the location of the PV park and the nearest five weather mini stations.

As we can see there are differences between the set of data that are published by each station and also the frequency data are read and stored in the station database. When we are considering the use of these data, we must take into consideration the appropriateness of this data to the purpose of our collection. Some data from one station can be more relevant for our study while for other data, integration of data from different station could be more relevant.

Data we are interested from the ministrations include temperature, humidity, precipitation level (hail and snow would receive a higher attention), solar radiation and pressure.

Historical data can be retrieved from the web page of the ministrations, while realtime data is collected using the similar method as for collecting data from the PV panel systems. We use appropriate scripts that call *phantomjs* software to browse and collect data from the webpage. Data collection must be synchronized with the specific frequency of data measurements of the station.

Care must be taken as the interface to the data can change and therefore data collecting process can be broken.

### 9.7.4.3    Cloud Services

Another category of data sources consists of sunrise and sunset calculators or a weather cloud service from where we can learn about the sunrise and sunset times for a specific location and for a specific day. This information is useful if we want to exclude data which is not relevant to our method (we need to restrict our data to the

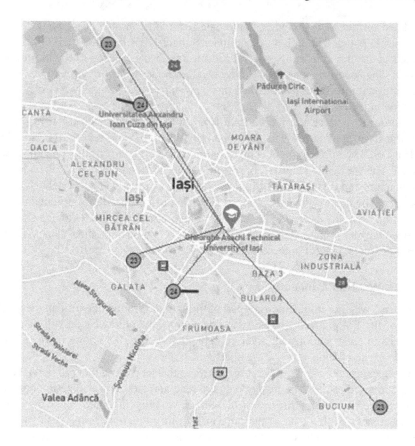

**Figure 9.9**   Location of PV platform and the closest weather mini stations

one corresponding to the daylight time). Of course, in other applications, it would be important to consider data generated during the night time and in other application we could be interested in the period of time around sunrise or sunset.

### 9.7.5   ML METHODOLOGY

Following the five main tasks in a common ML workflow we have used the modular ML pipeline shown in Figure 9.10.

The pipeline contains the usual blocks used in ML systems, this modularity having the advantage of being easily reused if new data sources are used or integrated with the initial ones.

### 9.7.5.1   Data Collection

In previous sections we have presented different data sources that can be used in the elaboration of the model used for monitoring PV panels. Each source has

**Figure 9.10**   The ML workflow

its characteristics and can be used for one or several function of the monitoring system.

In the following we will present a methodology that is using three of the presented sources:

> data collected from the solar PV platform using the application described in section 9.7.4;
> data collected from three closest weather mini stations part of the Weather Underground network (i.e. IAII19, IAII20 and IISVLADI2 Table 9.1. Integration of data from three weather mini stations was necessary as no single station was able to provide the entire set of data needed for model elaboration or the data publishing frequency was not high enough with respect to the other sources;
> data from a astronomical calculator for determining the sunrise and sunset times for the location of the PV park in the days considered for data set collection.

Figure 9.11 depicts the variation of some of the parameters collected from one of the MATE device associated with the PV panels part of the platform.

In a first stage of our study, data was collected and solar PV panels were observed during December 2018 and January 2019. PV panels observation was performed with the aim of elaborating a binary regression model. Therefore, during the observation there were noted if PV panels were covered or not covered with snow.

After analyzing the results from the first stage, it was concluded that during winter days PV can transition from the status covered to status uncovered in larger and finer number of steps. A second stage of the study was planned for winter 2019–2020 with the aim to record this finer transition between the two states. Unfortunately the collaborated factors of a dry winter and non functionality of the system during the few snowing days prevented us to collect necessary data.

In order to be able to go further with the development of this methodology, we resorted to the application of a method based on the use of hybrid data sources. In this sense, we used a PV panel model in Matlab and we fed the model with data taken from real data set sources.

The model we used is based on the Partial Shading of a PV Module model from Mathworks [31]. The model tries to simulate the situation when some of the cells of the PV panel are shaded making the bypass diodes to open and allow the current to

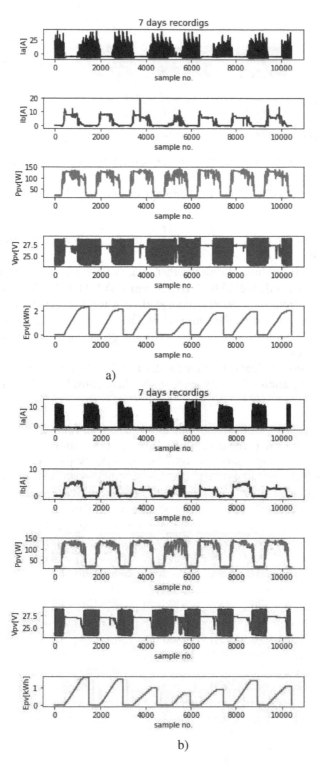

**Figure 9.11**   Solar panel parameters variation for a 7 day period a) single axis b) double axis.

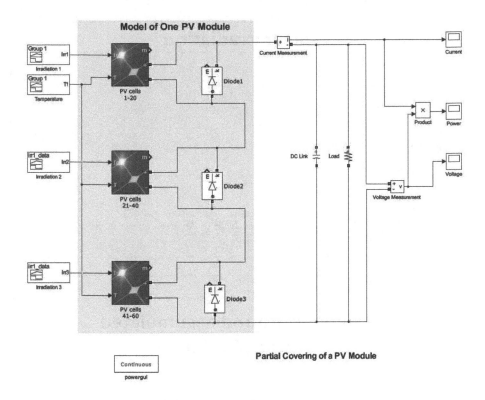

**Figure 9.12**  Simulink model for partial covering with snow

bypass these cells preventing their overheating or damaging. The modified model is given in Figure 9.12. The model has as inputs the solar irradiation and temperature and these are feed from the real data sources. As we can see the model considers three bypass diodes, one for each three groups of 20 PV cells (for a total of 60 PV cells per panel). Partial covering of the PV panel with snow is simulated by modifying the irradiation of the corresponding group of cells. Data for solar irradiation and temperature are input in the model using the signal builder block that takes data provided in csv files. For the PV cells considered covered with snow, an attenuation factor was used taking into account that snow has a high albedo (usually between 70% and 90%). Things are more complex as in [34] is shown that snow albedo can change not only with temperature increase that result in snow grains getting bigger, but also with the level of cloudiness and presence of other particles (dust, soot). Further researches will try to establish if a more precise modeling is necessary, for example taking into account also a snow melting model such as in [29].

Figure 9.13 presents data collected from weather ministrations (irradiation, temperature and precipitation) that are used to feed the model of PV panels partially covered with snow.

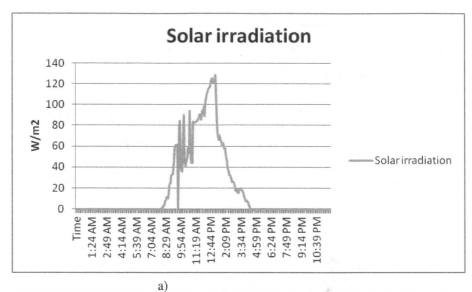

a)

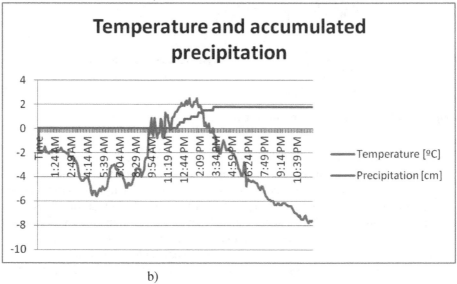

b)

**Figure 9.13**   Some data collected from weather mini stations on a snowing day in January 2019, a) solar irradiation, b) temperature and precipitation

### 9.7.5.2   Data Preprocessing

In order to ensure data quality and taking into account that data is collected from different sources, a pre-processing step was performed with the following objectives:

handling missing data due to communication errors or small periods of interrupted communication between the collecting system and data sources;
size reduction of data set;
aligning data to a unique daily timestamp frame as data collection or recording can be affected by the server executing other tasks;
synchronization of different data sources taking into account different frequencies used for measurement or publishing data;
elimination of time frames outside the interest range (e.g. night time);
filtering the measurement noise;
encoding variables.

### 9.7.5.3   Model Selection

The research presented in this chapter extends the results given in [17] where the PV panels have been observed with respect to their coverage with snow in the period December 2018 and January 2019. The ML techniques are applied in order to predict status (covered or not) of panels in a PV park based on data collected from the three different sources presented in Data Collection section. In our work we have tested different classification techniques i.e. logistic regression (LR), k near neighbors (KNN), support vector machine (SVM) and random forest (RF). We have applied these techniques on the same set of data, using the same training/test split percentage and compared the accuracy of prediction of the trained models when applied to the test set.

### 9.7.5.4   Feature Selection

By accessing the control and conversion system of the PV panels using the MATE devices we can read more than 35 parameters corresponding to instantaneous values of currents and voltages in different circuits of the conversion system, generated power and accumulated values of current or flowing energy through the system. In addition, we can collect system status data such as "Charging", "Inverting", "AC Use", "AC Drop" etc. Studies in literature and work in [17] shows that prediction is improved if we can select a good set of parameters to be used in the training process.

Table 9.4, reproduced in this work from [17] for comparison purposes, shows the result for selecting different parameter set.

### 9.7.5.5   Training and Validation

Training and validation of the model was performed using Python and machine learning algorithms implemented in the *sklearn* library [36]. The development environment used to code the ML algorithm was the Python Notebook.

**Table 9.4**

**Variation of Score with Feature Set for Logistic Regression Method**

| Set # | Feature 1 | Feature 2 | Feature 3 | Score | Rank |
|---|---|---|---|---|---|
| 1 | Channel B Current | PV Voltage | Solar Radiation | 0.916 | 2 |
| 2 | Channel A Current | Channel B Current | Solar Radiation | 0.846 | 5 |
| 3 | Channel A Current | Channel B Current | PV Voltage | 0.927 | 1 |
| 4 | Panel Output Current | Charged Energy | Solar Radiation | 0.831 | 7 |
| 5 | Panel Output Voltage | Battery Voltage | Solar Radiation | 0.890 | 3 |
| 6 | Inverter Input Voltage | Inverter Output Voltage | Solar Radiation | 0.857 | 4 |
| 7 | PV Voltage | Panel Output Current | Solar Radiation | 0.838 | 6 |

**Table 9.5**

**Variation of Score with Feature Set for K Near Neighbors Method**

| Set # | Feature 1 | Feature 2 | Feature 3 | Score | | Rank | |
|---|---|---|---|---|---|---|---|
| | | | | k = 3 | k = 5 | k = 3 | k = 5 |
| 1 | Channel B Current | PV Voltage | Solar Radiation | 0.956 | 0.898 | 1 | 2 |
| 2 | Channel A Current | Channel B Current | Solar Radiation | 0.826 | 0.811 | 4 | 4 |
| 3 | Channel A Current | Channel B Current | PV Voltage | 0.956 | 0.913 | 1 | 1 |
| 4 | Panel Output Current | Charged Energy | Solar Radiation | 0.797 | 0.739 | 5 | 5 |
| 5 | Panel Output Voltage | Battery Voltage | Solar Radiation | 0.942 | 0.898 | 2 | 2 |
| 6 | Inverter Input Voltage | Inverter Output Voltage | Solar Radiation | 0.855 | 0.855 | 3 | 3 |
| 7 | PV Voltage | Panel Output Current | Solar Radiation | 0.956 | 0.898 | 1 | 2 |

The initial data set was split in two subsets to be used for training and validation step respectively. Different split percentages were tested in order to see which one gives the best result.

In the case of logistic regression method the classifier model was selected and *linear* and *lbfgs* solvers were used. In [17], it was shown that using *lbfgs* solver has given better results.

The KNN method was applied in two cases, i.e. for k = 3 and for k = 5. For comparison purpose, the method was applied for the same different set of features used in the case of logistic regression. The results are given in Table 9.5.

The third method applied was *Support Vector Machine* with different kernels. The linear kernel proved to give the best results that are shown in Table 9.6.

The results obtained when applying the *Random Forest* method is given in Table 9.7 for the cases where number of trees were 10 and 100, respectively.

Analyzing Tables 9.4–9.7 and Figure 9.14 we can see that RF with number of trees equal to 100 gives the best results for predicting status of the PV panel. On the second position as method is the KNN with its variant where k = 3 performing better than for k = 5.

**Table 9.6**

**Variation of Score with Feature Set for Support Vector Machine Method**

| Set # | Feature 1 | Feature 2 | Feature 3 | Score | Rank |
|---|---|---|---|---|---|
| 1 | Channel B Current | PV Voltage | Solar Radiation | 0.927 | 1 |
| 2 | Channel A Current | Channel B Current | Solar Radiation | 0.869 | 4 |
| 3 | Channel A Current | Channel B Current | PV Voltage | 0.913 | 2 |
| 4 | Panel Output Current | Charged Energy | Solar Radiation | 0.884 | 5 |
| 5 | Panel Output Voltage | Battery Voltage | Solar Radiation | 0.898 | 3 |
| 6 | Inverter Input Voltage | Inverter Output Voltage | Solar Radiation | 0.811 | 6 |
| 7 | PV Voltage | Panel Output Current | Solar Radiation | 0.927 | 1 |

**Table 9.7**

**Variation of Score with Feature Set for Random Forest Method**

| Set # | Feature 1 | Feature 2 | Feature 3 | Score # of trees 10 | 100 | Rank # of trees 10 | 100 |
|---|---|---|---|---|---|---|---|
| 1 | Channel B Current | PV Voltage | Solar Radiation | 0.956 | 0.985 | 2 | 1 |
| 2 | Channel A Current | Channel B Current | Solar Radiation | 0.927 | 0.927 | 3 | 4 |
| 3 | Channel A Current | Channel B Current | PV Voltage | 0.971 | 0.985 | 1 | 1 |
| 4 | Panel Output Current | Charged Energy | Solar Radiation | 0.956 | 0.956 | 2 | 3 |
| 5 | Panel Output Voltage | Battery Voltage | Solar Radiation | 0.971 | 0.971 | 1 | 2 |
| 6 | Inverter Input Voltage | Inverter Output Voltage | Solar Radiation | 0.782 | 0.797 | 4 | 5 |
| 7 | PV Voltage | Panel Output Current | Solar Radiation | 0.956 | 0.956 | 2 | 3 |

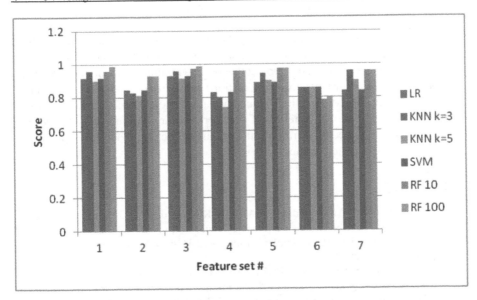

**Figure 9.14**   Comparison between different ML methods and for the same data and feature sets

For all methods, the third feature set is the set that gives the best results, followed closely by the first feature set. Future work will seek to apply the methods for an extended data set where a good candidate for a relevant (for snow coverage) feature would be the information about the level of precipitation.

The main purpose of this research was to start the initial development of a monitoring system and a method to be used in this system for the identification of possible degradation of PV panel operation based on machine learning (ML) techniques.

The results presented in this study are influenced by several factors among we can mention the fact that for the moment the PV park is not working at it full capacity and its connection to the grid is not finalized. The PV park has a small load consisting in an architectural lighting system that is connected only during the night time.

Further research will seek to extend the method to production PV parks and using data sets closer to a real situation.

## 9.8    CONCLUSIONS

In this work we have presented a method for monitoring and identifying the situation when solar panels in a PV platform are in a reduced operation status. This can be the result of coverage of the PV panel surface with different substances such snow, mud or sand. The use case presented takes into consideration coverage with snow, but the method can be easily extended to the case of other materials.

These particular conditions are specific to certain geographical locations (proximity to deserts, seas, or oceans), to some periods of the year (heavy storms, fog, snowing) or to specific temporary conditions of the PV system location (period of expansion of residential or industrial constructions or execution of heavy infrastructure works).

For developing this method we have applied some well known classification algorithms on a common dataset created with data collected from the Enered PV platform (which is part of a larger research project that deals with improvement of energy generation and consumption), weather mini stations and astronomical data calculators.

Early identification of situations when PV panels operation is degraded is aimed to be part of a monitoring system that can alert and are can promptly trigger the appropriate activities needed to remove the negative effects and to restore operational parameters to their  typical values.

## REFERENCES

1. Ahlam Althobaiti, Anish Jindal, and Angelos K Marnerides. SCADA-agnostic power modelling for distributed renewable energy sources. In *2020 IEEE 21st International Symposium on "A World of Wireless, Mobile and Multimedia Networks" (WoWMoM)*, pages 379–384, 2020.
2. Gagangeet Singh Aujla, Anish Jindal, Neeraj Kumar, and Mukesh Singh. SDN-based data center energy management system using res and electric vehicles. In *2016 IEEE Global Communications Conference (GLOBECOM)*, pages 1–6, 2016.

3. Gagangeet Singh Aujla, Neeraj Kumar, Mukesh Singh, and Albert Y Zomaya. Energy trading with dynamic pricing for electric vehicles in a smart city environment. *Journal of Parallel and Distributed Computing*, 127:169–183, 2019.

4. Gagangeet Singh Aujla, Amritpal Singh, and Neeraj Kumar. Adaptflow: Adaptive flow forwarding scheme for software defined industrial networks. *IEEE Internet of Things Journal, 7(7)*, pp. 5843–5851, 2019.

5. Gagangeet Singh Aujla, Mukesh Singh, Neeraj Kumar, and Albert Zomaya. Stackelberg game for energy-aware resource allocation to sustain data centers using res. *IEEE Transactions on Cloud Computing, 7(4)*, pp. 1109–1123, 2017.

6. T Babasaki and Y Higuchi. Using PV string data to diagnose failure of solar panels in a solar power plant. In *2018 IEEE International Telecommunications Energy Conference (INTELEC)*, pages 1–4, 2018.

7. Milan Belik. PV panels under lightning conditions. In *Proceedings of the 2014 15th International Scientific Conference on Electric Power Engineering, EPE 2014*, pages 367–370, 2014.

8. Samuel Bimenyimana, Godwin Norense Osarumwense Asemota, Mesa Cicilia Kemunto, and Lingling Li. Shading effects in photovoltaic modules: Simulation and experimental results. In *2017 2nd International Conference on Power and Renewable Energy, ICPRE 2017*, pages 904–909, Jun 2018.

9. Liviu Breniuc and Cristian Gyozo Haba. Embedded system for increasing home comfort and security. In *EPE 2014 – Proceedings of the 2014 International Conference and Exposition on Electrical and Power Engineering*, pages 881–886, Dec 2014.

10. M Catelani, L Ciani, L Cristaldi, M Faifer, M Lazzaroni, and M Rossi. Characterization of photovoltaic panels: The effects of dust. In *2012 IEEE International Instrumentation and Measurement Technology Conference Proceedings*, pages 1–4, 2012.

11. Deger Energie. DEGERtraker 5000NT Dual Axis Tracking System, Data Sheet, 2011. URL: https://www.degerenergie.de/deger-5000nt/

12. Deger Energie. DEGER TOPtracker 40NT Single Axis Tracking System, Technical Data, 2013. URL: http://planete-solar.fr/WordPress3/wp-content/uploads/2013/10/DEGER-TOPtracker-8.540NT-6.1_AS.pdf

13. Mohammad S. El-Shobokshy, and Fahmy M. Hussein. Degradation of photovoltaic cell performance due to dust deposition on to its surface. *Renewable Energy*, 3(6–7):585–590, Sep 1993.

14. EurObserv'ER. Photovoltaic Barometer. Technical report, DurObserv'ER, 2020.

15. V C Ferreira, R C Carrano, J O Silva, C V N Albuquerque, D C Muchaluat-Saade, and D Passos. Fault detection and diagnosis for solar-powered Wireless Mesh Networks using machine learning. In *2017 IFIP/IEEE Symposium on Integrated Network and Service Management (IM)*, pages 456–462, 2017.

16. Cristian Gvozo. Haba. Monitoring photovoltaic parks for damage prevention and optimal operation. In *2017 International Conference on Electromechanical and Power Systems (SIELMEN)*, Jan 2017.

17. Cristian Gvozo Haba. Monitoring solar panels using machine learning techniques. In *Proceedings of 2019 8th International Conference on Modern Power Systems, MPS 2019*, May 2019.

18. IRENA. Future of Solar Photovoltaic: Deployment, investment, technology, grid integration and socio-economic aspects. (A Global Energy Transformation: paper). Technical report, International Renewable Energy Agency, Abu Dhabi., 2019.

19. F Jawaid and K NazirJunejo. Predicting daily mean solar power using machine learning regression techniques. In *2016 6th International Conference on Innovative Computing Technology (INTECH)*, pages 355–360, 2016.

20. Anish Jindal, Gagangeet Singh Aujla, Neeraj Kumar, Radu Prodan, and Mohammad S Obaidat. Drums: Demand response management in a smart city using deep learning and SVR. In *2018 IEEE Global Communications Conference (GLOBECOM)*, pages 1–6, 2018.

21. Anish Jindal, Bharat Singh Bhambhu, Mukesh Singh, Neeraj Kumar, and Kshirasagar Naik. A heuristic-based appliance scheduling scheme for smart homes. *IEEE Transactions on Industrial Informatics*, 16(5):3242–3255, 2019.

22. Anish Jindal, Neeraj Kumar, and Mukesh Singh. A unified framework for big data acquisition, storage, and analytics for demand response management in smart cities. *Future Generation Computer Systems*, 108:921–934, 2020.

23. Anish Jindal, Angelos K Marnerides, Andrew Scott, and David Hutchison. Identifying security challenges in renewable energy systems: A wind turbine case study. In *Proceedings of the Tenth ACM International Conference on Future Energy Systems*, pages 370–372, 2019.

24. Anish Jindal, Alberto Schaeffer-Filho, Angelos K Marnerides, Paul Smith, Andreas Mauthe, and Lisandro Granville. Tackling energy theft in smart grids through data-driven analysis. In *2020 International Conference on Computing, Networking and Communications (ICNC)*, pages 410–414, 2020.

25. Dirk C Jordan and Sarah R Kurtz. Photovoltaic Degradation Rates – An Analytical Review: Preprint. Technical report, 2012.

26. J. K. Kaldellis and M. Kapsali. Simulating the dust effect on the energy performance of photovoltaic generators based on experimental measurements. *Energy*, 36(8):5154–5161, Aug 2011.

27. Neeraj Kumar, Gagangeet Singh Aujla, Sahil Garg, Kuljeet Kaur, Rajiv Ranjan, and Saurabh Kumar Garg. Renewable energy-based multi-indexed job classification and container management scheme for sustainability of cloud data centers. *IEEE Transactions on Industrial Informatics*, 15(5):2947–2957, 2018.

28. Neeraj Kumar, Tanya Dhand, Anish Jindal, Gagangeet Singh Aujla, Haotong Cao, and Longxiang Yang. An edge-fog computing framework for cloud of things in vehicle to grid environment. In *2020 IEEE 21st International Symposium on "A World of Wireless, Mobile and Multimedia Networks" (WoWMoM)*, pages 354–359, 2020.

29. Xiaobing Liu, Simon J Rees, and Jeffrey D Spitler. Modeling snow melting on heated pavement surfaces. Part I: Model development. *Applied Thermal Engineering*, 27(5):1115–1124, 2007.

30. Dragos Machidon, Roxana Oprea, and Marcel Istrate. Considerations on the Opportunity of Using Various Optimum Tilt Angles for Fixed Photovoltaic Panels in Iaşi. *Buletinul Institutului Politehnic din Iaşi*, 69(3):79–92, 2019.

31. Mathworks. Partial shading of a PV module. URL: https://www.mathworks.com/help/ physmod/ sps/ ug/partial-shading-of-a-pv-module.html

32. M K Mazumder, M N Horenstein, C Heiling, J W Stark, A Sayyah, J Yellowhair, and A Raychowdhury. Environmental degradation of the optical surface of PV modules and solar mirrors by soiling and high RH and mitigation methods for minimizing energy yield losses. In *2015 IEEE 42nd Photovoltaic Specialist Conference (PVSC)*, pages 1–6, 2015.

33. V Mehra, R Ram, and C Vergara. A novel application of machine learning techniques for activity-based load disaggregation in rural off-grid, isolated solar systems. In *2016 IEEE Global Humanitarian Technology Conference (GHTC)*, pages 372–378, 2016.

34. Peter Kuipers Munneke. *Snow, Ice and Solar Radiation*. PhD thesis, Institute for Marine and Atmospheric Research, Utrecht, Netherlands, 2009.

35. K Niazi, W Akhtar, H A Khan, S Sohaib, and A K Nasir. Binary classification of defective solar PV Modules using thermography. In *2018 IEEE 7th World Conference on Photovoltaic Energy Conversion (WCPEC) (A Joint Conference of 45th IEEE PVSC, 28th PVSEC & 34th EU PVSEC)*, pages 753–757, 2018.

36. F Pedregosa, G Varoquaux, A Gramfort, V Michel, B Thirion, O Grisel, M Blondel, P Prettenhofer, R Weiss, V Dubourg, J Vanderplas, A Passos, D Cournapeau, M Brucher, M Perrot, and E Duchesnay. Scikit-learn: Machine learning in Python. *Journal of Machine Learning Research*, 12:2825–2830, 2011.

37. Contero Salvadores and Jose Francisco. *Shadowing Effect on the Performance in Solar PV-Cells*. PhD thesis, University of Gavle, Sweden, 2015.

38. Mariusz T. Sarniak, Jacek Wernik, and Krzysztof J. Wołosz. Application of the double diode model of photovoltaic cells for simulation studies on the impact of partial shading of silicon photovoltaic modules on the waveforms of their current–voltage characteristic. *Energies*, 12(12):2421, Jun 2019.

39. N N B Ulapane and S G Abeyratne. Gaussian process for learning solar panel maximum power point characteristics as functions of environmental conditions. In *2014 9th IEEE Conference on Industrial Electronics and Applications*, pages 1756–1761, 2014.

40. Chengqing Yuan, Conglin Dong, Liangliang Zhao, and X Yan. Marine environmental damage effects of solar cell panel. In *2010 Prognostics and System Health Management Conference*, pages 1–5, 2010.

41. X Zhang, R Fan, W Li, S Wei, G Du, L Zhang, and H Wei. Solar photovoltaic array surface cleaning device and control means. In *2016 Chinese Control and Decision Conference (CCDC)*, pages 5982–5983, 2016.

42. Rodrigues, S., Ramos, H. G., & Morgado-Dias, F. (2018). Machine learning PV system performance analyser. Progress in Photovoltaics: *Research and Applications*, 26(8), 675–687.

43. Howlader, M. M., Howlader, M. M., Rokonuzzaman, M., Khan, M. S. A., Nur, A. U., & Al Amin, A. (2017, September). GIS-based solar irradiation forecasting using support vector regression and investigations of technical constraints for PV deployment in Bangladesh. In *2017 4th International Conference on Advances in Electrical Engineering (ICAEE)* (pp. 675–680). IEEE.

44. Martin, R., Aler, R., Valls, J. M., & Galván, I. M. (2016). Machine learning techniques for daily solar energy prediction and interpolation using numerical weather models. *Concurrency and Computation: Practice and Experience*, 28(4), 1261–1274.

45. Su, D., Batzelis, E., & Pal, B. (2019, September). Machine learning algorithms in forecasting of photovoltaic power generation. In *2019 International Conference on Smart Energy Systems and Technologies (SEST)* (pp. 1–6). IEEE.

46. Ferreira, V. C., Carrano, R. C., Silva, J. O., Albuquerque, C. V., Muchaluat-Saade, D. C., & Passos, D. (2017, May). Fault detection and diagnosis for solar-powered wireless mesh networks using machine learning. In *2017 IFIP/IEEE Symposium on Integrated Network and Service Management (IM)* (pp. 456–462). IEEE.

47. Shao, X., Lu, S., van Kessel, T. G., Hamann, H. F., Daehler, L., Cwagenberg, J., & Li, A. (2016, December). Solar irradiance forecasting by machine learning for solar car races. In *2016 IEEE International Conference on Big Data (Big Data)* (pp. 2209–2216).

48. Enache, B. A., Bîrleanu, F. M., & Răduṭ, M. (2016, June). Modeling a PV panel using the manufacturer data and a hybrid adaptive method. In *2016 8th International Conference on Electronics, Computers and Artificial Intelligence (ECAI)* (pp. 1–6). IEEE.

49. Sun, Y., Venugopal, V., & Brandt, A. R. (2018, June). Convolutional neural network for short-term solar panel output prediction. In *2018 IEEE 7th World Conference on Photovoltaic Energy Conversion (WCPEC)(A Joint Conference of 45th IEEE PVSC, 28th PVSEC & 34th EU PVSEC)* (pp. 2357–2361).

50. Tao, Y., Zhang, M., & Parsons, M. (2017, July). Deep learning in photovoltaic penetration classification. In *2017 IEEE Power & Energy Society General Meeting* (pp. 1–5).

51. Paridari, K., Azuatalam, D., Chapman, A. C., Verbič, G., & Nordström, L. (2018, October). A plug-and-play home energy management algorithm using optimization and machine learning techniques. In *2018 IEEE International Conference on Communications, Control, and Computing Technologies for Smart Grids (SmartGridComm)* (pp. 1–6).

# 10 Intelligent Control System for Smart Environment Using Internet of Things

*Chintan Bhatt*
Charotar University of Science and Technology, Changa, Anand, India

*Riya Patel*
Charotar University of Science and Technology, Changa, Anand, India

*Siddharth Patel*
Charotar University of Science and Technology, Changa, Anand, India

*Hussain Sadikot*
Charotar University of Science and Technology, Changa, Anand, India

*Akrit Khanna*
Charotar University of Science and Technology, Changa, Anand, India

*Esha Shah*
Charotar University of Science and Technology, Changa, Anand, India

## CONTENTS

## 10.1  INTRODUCTION

The concept of the *Internet of Energy (IoE)* [1] is derived from the concept of the *Internet of Things (IoT)* [2, 3], which uses the Internet to access and connect appliances. The concept behind the term IoE was to create more efficient and automatic systems that specifically dealt with appliances like bulbs, fans, and other energy appliances which when connected using a system, and developed using principles of IoE helps in minimizing energy consumption and wastage of energy using the systematic infrastructure of appliances. IoE based infrastructures mainly have generation, transmission, and efficient usage of energy.

Energy consumption in heating, ventilation and air conditioning (HVAC) typically accounts for half of energy consumption in buildings [4]. If we could think of some way, using which we can minimize the energy that gets wasted in our daily routine activities then, we can individually save an ample amount of energy on a daily basis. This could not only save the light units but also lead to taking a step towards finding innovative and efficient solutions for harvesting energy for our own good.

IoT has played a multidisciplinary role in the development of *smart devices* and *control systems* substituting the use of wired systems with the computer interfaced system alternatives in many industries. Home Automation [5, 6] is an excellent example of the ubiquitous nature of the application of IoT where interfaces like buttons and switches in normal houses are substituted by a central control system allowing easy access of devices with smartphones. IoT can be regarded as a system constituting devices in the real world, and sensors [7] attached or combined to these devices and connected to the Internet via wired and wireless network structure. Several types of connections and standards can be used for IoT sensors like Bluetooth, Wi-Fi, and ZigBee with technologies like GSM, GPRS, 3G, and LTE to provide wide area connectivity [8, 9].

Computer Vision is another interdisciplinary scientific field that adds *intelligence* to such control systems. It relates to providing computers with an understanding of digital images or videos.

We can utilize the power of these technologies by combining them together to have complete control over the connected appliances and *minimize* the power consumption by detecting the presence of people and turning ON or OFF the appliances accordingly. One important question that could arise while managing the appliances remotely is managing the access to those appliances when implementing in a building or an organization, where a large number of people are involved and many of them could have access to many locations within the organization.

Providing privileged access to each user on the basis of his/her privilege level is the call. This project provides the prototype solution for implementing remote access with privileged access-rights along with automatic control. The users can have access with the help of the mobile application where manual control has the *priority* over automatic control.

The mobile application for our prototype is developed using the Flutter framework so as to have cross-platform support and have applications for both Android as well as iOS. The prototype is built taking into consideration the scenario of a college/university with three privilege levels as:

*Super-admin*: has privilege to create or delete users and locations,

*Admins*: can manage permissions to allotted locations by granting and revoking access to and from the Students (or employees in an organization).

*Students (or employees in an organization)*: can only access appliances of locations he/she is permitted to by Admin(s).

The higher class has all the privileges as its lower class but not vice versa. The users can check the *live* status of the appliances at permitted locations. This is how strict monitoring and managing permissions can be done in a big organization (here, a college/university) and it can also avoid any misbehavior. Most importantly, it would result in an effective minimization of the power consumption that usually occurs due to the appliances left ON until someone encounters this and turns it OFF. It can also turn out to be helpful when one wishes to put ON appliances such as an air conditioner a few minutes before going to that location by remotely accessing from where he/she is.

## 10.2   RELATED WORK

Wireless Home Automation System (WHAS) involves communications, control, and information processing across various systems all done over the internet using IoT. To control basic home appliances remotely via computer or mobile devices, home automation systems using ZigBee use wireless sensors between the controller and user section [10]. The transmission process involves the input fed from mobile or PC through a web application and the receiver section ZigBee RX receiving the information processes the required output. The microcontroller is implemented in WHAS and once the wireless connection is created, the system implements a transmitter-receiver to establish a ZigBee connection.

Controlling raspberry pi GPIO wirelessly [12] can be done using Bluetooth or a web server. But connecting it with Bluetooth creates a problem (an issue) to control the device remotely, thus to make a connection remote, one can create their own web server, which will control the GPIO with the help of HTTP requests. The simplest way to create a web server is to use the Flask library. So, whenever a request is made, a block of code is executed, and that block contains the code to control the specific GPIO pin to control it remotely.

An IoT based surveillance and control system for devices using ESP module and Wi-Fi [13] can be used to continuously track the state, i.e. ON or OFF, and working condition of electrical devices by operating them from a central remote location. It grants access and manages both indoor and outdoor lighting. It checks for defects in the devices using sensors and sends an alert using analog signals, on encountering a defect to make the system more efficient. In order to reduce or minimize the

**Table 10.1**

**Comparison between Different Wireless Technologies**

| Technology | Range | Data Rate | Power Usage (Battery Life) | Security | Installation Cost |
|---|---|---|---|---|---|
| Wi-Fi (802.11n) | ≤ 70 m | 11 Mbps | High (Hours) | High | High |
| Bluetooth | ≤ 50 m | 1 Mbps | Low (Few Months) | High | Low |
| Zigbee | ≤ 100 m | 250 Kbps | Very Low (5–10 years) | Low | Low |

consumption of energy, it provides central control over the electrical appliances for street lighting systems.

The electrical appliances at home can be connected to Internet Controlled Switch Box [14] to enable controlling them remotely at any time. The smart switch box integrates the power supply unit, Wi-Fi switch, relay, two-way switch, and an Android application on the mobile device. ESP8266 chip acts as the Wi-Fi device with a web socket implemented in it which connects the smart switch box with the mobile phone in the same Wi-Fi environment. To turn an appliance ON or OFF, connected with the smart box, the design chooses ESP8266 as the main control chip to control the relay. Table 10.1 shows the comparison between different wireless technologies used to implement smart systems.

The electrical appliances connected with a remote server can be controlled from a remote location using Computer Interfacing [15] by logging in to the server PC. To log in, a web page is created where real IP is used or Microsoft remote assistance system where virtual IP is used. Once logged in, the client or end-user can view the live status of any connected appliance as well as get status updates for those appliances via the developed application [16]. The data, consisting of the id number of the client PC shows the status of the electrical appliance within an instant, is accepted by TCP/IP listener, and then sent to a specific port of the desired address over the internet by using the device.

The REST architecture style was designed specifically for working with components such as media components, files, or even objects on a particular hardware device. A restful service [17] would use the normal HTTP verbs of *get*, *post*, *put*, and *delete* for working with the required components. *Get* request is used for retrieving the data, *post* request is used to insert data whereas *put* request is used to insert and update data, and *delete* request is used for deleting the data respectively. REST permits different data formats such as Plain text, HTML, XML, JSON, etc. of which the most preferred format for transferring data is JSON.

Haar cascade [18] is a sequence of square shaped functions which forms a family of wavelets or base. Model is trained on positive and negative dataset. Then the features are identified by the haar cascades. Features are identified by reiterating features over an Image. Here, a large amount of computation, power and time is consumed.

## 10.3   METHODOLOGY

This project aims to control the electrical appliances like fans and LEDs inside the lab/room/office considering the problem of access control and it's solutions. Before getting into the implementation details, let's ponder into the details of the components used.

*Raspberry Pi* - It is a very basic and cost-effective computer that runs on Linux OS, and provides a set of GPIO (general purpose input/output) pins that allows one to control electronic components for physical computing, and explore the Internet of Things (IoT).

*Relay Module* - It is a separate hardware device used for remote device switching. It helps to remotely control devices over a network or the Internet. Devices can be remotely powered ON or OFF with commands coming from the raspberry pi. It helps to control computers, peripherals or other powered devices from across the office or across the world.

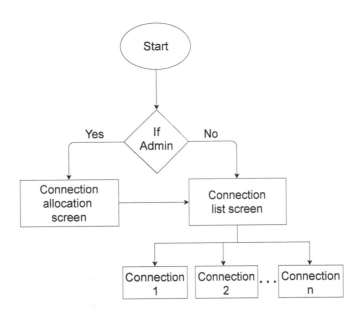

**Figure 10.1**   Privileged access control

### 10.3.1   PRIVILEGED ACCESS

If any user having the application can control any of the appliances remotely of the lab/room/office, it might cause a nuisance if the user misuses it. To overcome this problem, the project has a privileged access control system. The approach is to *grant* and *revoke* access permissions to only selected users by the user of higher privilege level (here, Admins) as and when needed. Django Rest Framework is used in the prototype to implement a *multi-user hierarchy* and permissions are managed according to the requirements, via the mobile application. Figure 10.1 shows the privileged access control mechanism.

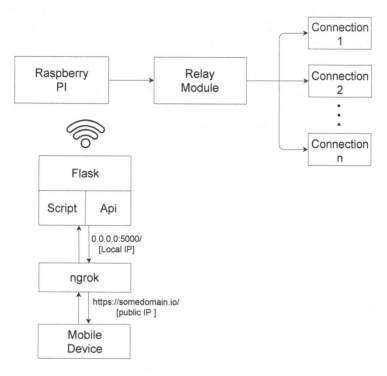

**Figure 10.2**   Manual access via mobile application

## 10.3.2   MANUAL CONTROL OF ELECTRICAL APPLIANCES

Figure 10.2 demonstrates the manual access via mobile application. To get control of the appliances, the prototype uses a relay module connected to Raspberry Pi's GPIO(s). There is a script to control the appliances via a mobile device and APIs for different operations are made using the Flask framework. The relay module signal pin is connected to the raspberry GPIO. Hence, when the API is called, the specific operation assigned to the pin is performed.

### 10.3.3    AUTOMATIC CONTROL OF ELECTRICAL APPLIANCES

Figure 10.3 demonstrates the automatic detection and control mechanism. Images are captured at a fixed interval of time from a camera device in the room and each image is sent to the server for processing. At the server-side, when the image is provided, it detects the presence of person(s) in it by dividing the image into different grid sections. The appliance in a particular section will turn ON if the presence of a person is detected in that section. This is implemented in the prototype using the OpenCV library for Image Processing. OpenCV is an open-source library that is mainly used to deal with problems and solutions related to Computer Vision applications and also helps to enhance the algorithms of Machine Learning. An API request is sent to the raspberry pi server with the corresponding pin number and hence, the state of the appliance is controlled.

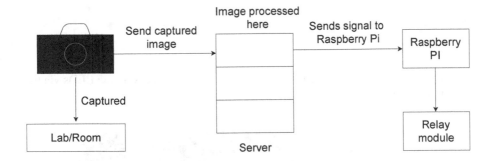

**Figure 10.3**    Automatic detection and control

### 10.4    EXPERIMENTAL SET-UP AND RESULTS

The relay module is connected to the raspberry pi through GPIO pins. These pins are used to give signals to relay modules. They consist of two signals, HIGH and LOW. When the signal is HIGH, the relay channel is turned ON and when the signal is LOW, the relay is turned OFF. In the prototype, a 4-channel relay module is used, i.e. up to four devices can be connected to one module at a time. Each device on the relay channel has a dedicated pin connected to raspberry pi's GPIO. The script is programmed according to the requirements and then the module is controlled by running that script. Figure 10.4 shows the circuit of the experimental set-up of the prototype.

Each relay has three sockets in which the negative pin of the LED is connected to the second socket, while the positive pin is connected to the power source (switchboard), and from the third socket a wire is connected to the negative of the power source (switchboard). Thus, this acts as a simple switch like the one in our home, but it is

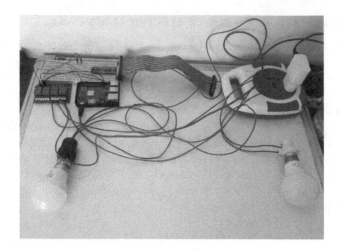

**Figure 10.4** Experimental set-up

operated using a python script. Figure 10.5 demonstrates accessing the LEDs via the Mobile application.

**Figure 10.5** Accessing LEDs via mobile application

On an average, the energy consumption by a single 8W LED bulb is approximately 8 Wh (0.008 KWh) and Corporate buildings or Educational buildings like schools or colleges, on average contain 10–12 LED bulbs per room. Assuming a room with 12 such LED bulbs and all being in use, the energy consumed is approximately 100 Wh (0.1 KWh).

It often happens that a person entering the room turns all the appliances ON in the first go irrespective of the part of the room he/she is going to sit and work. In ideal conditions, only 1 or 2 LED bulbs (consuming $\approx$ 16 Wh energy) should be lit for a single person, but in an actual scenario, all LED bulbs are lit. This leads to the unnecessary energy consumption of almost 80 Wh (0.08 KWh) for the unnecessarily lit 10 LED bulbs.

By using the proposed smart control system, the issue of energy consumption can be brought down to a minimum by intelligently controlling the number of LEDs or appliances that are turned ON. Using the computer vision module, the system will automatically detect the presence of people in that room and accordingly turn ON the appliances where the person is. Figure 10.6 shows the relation between Energy consumption with the number of people in the room by using the smart system.

According to model training conducted on the Computer Vision Model of the prototype, the average accuracy of the Model is found to be about 50%. Using this data, we can predict that the energy consumption can be brought to a minimum by as much as 40 Wh.

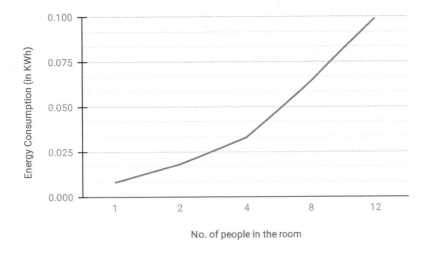

**Figure 10.6**   Energy consumption using smart system

Average Energy Consumption by conventional system = 100 Wh

Average Energy Consumption using Smart System = 60 Wh

Average Energy Saved per hour using Smart System = 100 – 60

$$= 40 \text{ Wh}$$

Total energy saved in a year by implementing Smart System $\approx 40*24*365$ Wh.

$$\approx 350.400 \text{ KWh.}$$

## 10.5   DISCUSSION AND CONCLUSION

Nowadays, the concept of minimization of consumption of energy is more important and getting more critical due to the lack of resources and an over-demand of energy. As a result, there is a surge in demand for energy, but only limited supply is possible due to limited resources.

Also due to some limitations and inefficiency to manage the energy consumption, the conventional light control systems are not feasible for the long run and they can't be successfully installed. The ease of access also plays a major factor in developing Smart Systems using which the consumers can easily access their appliances remotely.

The proposed system minimizes energy consumption as well as gives the consumer remote access to their appliances. Energy is saved using the trained model of Computer Vision which accurately identifies how much light intensity is required in a particular room of the building and manages it automatically in order to save the energy with minimum interaction and also the system provides privilege access to its users, i.e. it provides the system the capability to allow different users to access same appliance where some users might have access to particular appliance temporarily while some may have it permanently. Thus, by using the system proposed, we can simultaneously achieve both the goals.

## 10.6   LIMITATIONS

The current prototype has some limitations which are now in the modification stage and are being worked upon for the development of the system. The primary limitation faced is that the HTTP response and request are slow when the internet connection is weak. This is due to the free version of RESTful APIs which has higher latency compared to paid APIs. Another issue faced is the presence detection of person(s) done using the Computer Vision module which is moderately accurate. A proper dataset to train the model of Computer Vision is not available, so a dataset is being developed using which more accurate and efficient model can be made and the main purpose of minimizing consumption of energy can be achieved as predicted theoretically while prototyping the system. The prototype is limited to light appliances only.

## 10.7   FUTURE ENHANCEMENTS

The system is under development using the customized dataset developed. Therefore, the proposed model is to develop a system more efficient than the current prototype which has an accuracy of around 50% only for the presence detection of person(s) in a room. Therefore, the Computer Vision module can be made more efficient by using better algorithms for presence detection in a room that is cost-efficient to optimize the consumption of energy.

The system facing the issue of high latency in HTTP requests and responses can be resolved by developing more efficient APIs. The system can be made compatible with more electrical appliances unlike the prototype currently compatible only with light appliances. The application of the system can be made available for both iOS and Android platforms with complete integration of all modules with higher efficiency.

## REFERENCES

1. H. Doost. 2018. Internet of Energy: A solution for improving the efficiency of reversible energy. *IEEE Global Engineering Education Conference (EDUCON)*, Tenerife, pp. 1890-1895, doi: 10.1109/EDUCON.2018.8363466.
2. A. Jindal, N. Kumar, M. Singh. 2020. Internet of Energy-Based Demand Response Management Scheme for Smart Homes and PHEVs Using SVM. *Future Generation Computer Systems,* vol. 108, pp. 1058-1068.
3. M. Bhayani, M. Patel, C. Bhatt. 2016. Internet of Things (IoT): In a Way of Smart World. *Advances in Intelligent Systems and Computing,* vol. 43 pp. 343-350, doi: 10.1007/978-981-10-0767-5_37.
4. V. Vakiloroaya, B. Samali, A. Fakhar, K. Pishghadam 2014. A Review of Different Strategies for HVAC Energy Saving. *Energy Conversion and Management,* vol. 77 pp. 738-754
5. N. Barodawala, B. Makwana, Y. Punjabi, C. Bhatt. 2018. Home Automation Using IoT. In: *Internet of Things and Big Data Analytics Toward Next-Generation Intelligence,* ed. Dey N., Hassanien A., Bhatt C., Ashour A., Satapathy S., 219-242. Studies in Big Data, vol 30. Cham, Springer.
6. A. Jindal, B. S. Bhambhu, M. Singh, N. Kumar, K. Naik. 2019. A Heuristic-Based Appliance Scheduling Scheme for Smart Homes. *IEEE Transactions on Industrial Informatics*, vol. 16, no. 5, pp. 3242-3255.
7. S. D. T. Kelly, N. K. Suryadevara and S. C. Mukhopadhyay. 2013. Towards the Implementation of IoT for Environmental Condition Monitoring in Homes. *IEEE Sensors Journal*, vol. 13, no. 10, pp. 3846-3853, doi: 10.1109/JSEN.2013.2263379.
8. Dr. K. Sailaja, Rohitha, M. 2018. Literature survey on real world applications using Internet of Things. *IADS International Conference on Computing, Communications & Data Engineering (CCODE)*, Available at SSRN: https://ssrn.com/abstract=3165327.
9. A. Jindal, A. K. Marnerides, A. Gouglidis, A. Mauthe, D. Hutchison. 2019. *Communication standards for distributed renewable energy sources integration in future electricity distribution networks.* In 2019-2019 IEEE International Conference on Acoustics, Speech and Signal Processing (ICASSP), pp. 8390-8393.
10. K. Karuppasamy, S. Gowtham, S. A. Athira, M. Ajitha. 2016. Controlling the Home Appliances Remotely through Web Application Using ZigBee. *International Journal for Research in Applied Science & Engineering Technology*, vol. 4, no. III (March) pp. 852-856.
11. Saleem, A., Khan, A., Malik, S. U. R., Pervaiz, H., Malik, H., Alam, M., Jindal, A. (2020). FESDA: Fog-Enabled Secure Data Aggregation in Smart Grid IoT Network. *IEEE Internet of Things Journal,* vol. 7, no. 7, pp. 6132 - 6142.

12. M. N. Azni, L. Vellasami, A. H. Zianal, et al. 2016. Home automation system with android application. *3rd International Conference on Electronic Design (ICED)*, Phuket, pp. 299-303, doi: 10.1109/ICED.2016.7804656.

13. A. K. Gupta and R. Johari. 2019. IOT based electrical device surveillance and control system. *4th International Conference on Internet of Things: Smart Innovation and Usages (IoT-SIU)*, Ghaziabad, pp. 1-5, doi: 10.1109/IoT-SIU.2019.8777342

14. Karthi, A. & Gs, Sabaresh. 2016. Design of Internet Controlled Switch Box using IoT. *Journal of Advances in Chemistry*, vol. 12, pp. 5104-5108.

15. N. Das, M. Alamgir, C. Das, M. Hasan, M. Rahman, and M. Alam. 2011. Developing A Smart Control System of Electrical Appliances By Computer Interfacing, *MIST,* vol. 3, p. 1.

16. S. Roy, S. Chatterjee, A. K. Das, S. Chattopadhyay, N. Kumar and A. V. Vasilakos. 2017. On the Design of Provably Secure Lightweight Remote User Authentication Scheme for Mobile Cloud Computing Services. *IEEE Access*, vol. 5, pp. 25808-25825, doi: 10.1109/ACCESS.2017.2764913.

17. Masse, M. 2011. *REST API Design Rulebook*, O'Reilly, Sebastopol.

18. M. Khan, S. Chakraborty, R. Astya and S. Khepra. 2019. Face detection and recognition using open CV. In *International Conference on Computing, Communication, and Intelligent Systems (ICCCIS),* Greater Noida. pp. 116-119, doi: 10.1109/ICCCIS48478.2019.8974493.

# 11 Pathway and Future of IoE in Smart Cities: Challenges of Big Data and Energy Sustainability

*Sharda Tripathi*
Department of Electronics and Telecommunications,
Politecnico di Torino, Turin, Italy

*Swades De*
Department of Electrical Engineering,
Indian Institute of Technology Delhi, New Delhi, India

## CONTENTS

## 11.1 INTRODUCTION

With the proliferation of miniature embedded intelligence, data-centric user applications as well as machine-type applications have been evolving at a fast pace. On one hand, the users' data demand as well as volume of data generated through multimedia-centric, virtual-reality/augmented-reality applications are expected to grow enormously. On the other hand, the interest of automated monitoring, control, and actuation of civic infrastructures and amenities is expected to cause a massive growth in deployment of telemetric devices. These trends have been pointing to new challenges, namely, (a) data handling capability of the communication networks [122], (b) imparting with appropriate intelligence at various stages of the network [5], and (c) green energy solutions and sustainability [38].

Typically, the Internet of Things (IoT) devices do not have sufficient data processing intelligence, whereas with various system health monitoring and commercial objectives the sampling rates are high [39]. Though each of such devices generate data on the order of a few tens of kbps, the scale of deployment of such telemetric devices in smart city, manufacturing, and medical diagnostic related applications is expected to be very high. Therefore, the net impact of the data generated by such devices on the communication infrastructure for their transmission can be very high [122]. Hence, application context aware processing of data at various stages of the network is imperative to keep the communication network from being overloaded. Heterogeneity of data, nonhomogeneous traffic load across the network, and non-ideal nature of communication channels and networks pose additional challenges. The real-life data being dynamic stochastic in nature and also the application contexts could widely vary [106]. These data context-driven features call for unique nature of learning based solutions.

While node-level and network intelligence are critical for reducing the load on resources at the node-level as well as network-level, large-scale deployment of IoT nodes over large and potentially inaccessible areas for massive machine-type communications (mMTC) can be feasible if the energy needs are addressed via green and more cost-efficient solutions [38]. Towards green energy solutions it is expected that the usage of ambient energy resources, such as solar, wind, ambient radio frequency (RF), will be judiciously exploited even if the energy from power grid is available. The power grid connectivity to energy hungry wireless stations, such as the clusterheads or edge nodes that connect the field IoT devices for telemetric operations, the Wi-Fi access points that provide mostly fixed-wireless connectivity, the cellular base stations that serve predominantly for mobile wireless connectivity, will act as the resources for two-way energy supply. While the power grid energy supply acts as a backup for energy to the wireless nodes, whenever possible, these wireless nodes also supply their surplus energy to the grid, thereby acting as microgrid power generators. This two-way energy flow in the smart grid act as the basis of Internet of Everything (IoE) [17]. There are several design optimization aspects for cost, i.e. capital expenditure (CAPEX) and operational expenditure (OPEX), minimization from the wireless network operation perspectives [97]. There are also issues pertaining to power grid network stability that need attention in the IoE [27]. While the smart power grid acts as a facilitator for the communication networks, power-grid stability assessment through its monitoring is associated with big data handling requirements and network-level challenges [28, 29]. Its role on integrating renewable energy along with the conventional sources in supplying power to the communication and services infrastructure is crucial.

This chapter highlights the application context-specific evolving needs for communication networks as well as the power grid networks, with the associated unique features and challenges, and the research trends to deal with them.

**Chapter organization:** In Section 11.2, the four broad categories on smart infrastructure: smart grid monitoring, advanced power metering, smart telemetry and actuation of civic infrastructure and amenities, and smart wireless services, are presented. Subsequently in Section 11.3, the associated big data challenges and massive machine-type communication needs are presented. Here, application context-specific unique learning-based solutions to deal with the node-level as well as network level are discussed. Specifically, light-weight machine learning solutions are presented to address the energy and communication bandwidth challenges. Further, network architecture level intelligent deployment for efficient resource usage are discussed. These include intelligent sensing and data pruning through fog computing, edge computing, as well as cloud computing. The adaptive solutions accounting jointly for data-level intelligence along with the intelligence on network constraints are discussed. Section 11.4 deals with the IoT node-level and network energy sustainability and IoE stability aspects, wherein the benefits from node-level and network intelligence (discussed in Section 11.3) are presented. Individual node-level energy harvesting features and the associated need for energy grid connectivity are presented. Network-level intelligence for self-energy-sustainability and two-way energy flow in IoE are presented. Here, an important aspect of power grid stability and controllability aspect is addressed. Cost-revenue tradeoff with smart services infrastructure are also discussed from game theory and economics perspectives. Finally, in Section 11.5 the book chapter is concluded.

## 11.2   IOE APPLICATION CASE STUDIES

In this section we present the various IoT and other communication applications that are linked to the evolving concepts of IoE. The IoT applications are predominantly for smart cities and smart infrastructure development, aiding to remote monitoring, control, and automation purposes. Examples of such applications include city/remote surveillance, pollution localization and control, mobile wireless services, automated billing, smart agriculture, and nature protection activities. A few typical application contexts are highlighted: (a) air/water quality monitoring and pollution localization, (b) smart grid health monitoring for efficient ways to grid stability control, (c) smart metering, (d) mobile agent assisted IoT for intelligent data exchange and energy replenishment. The common denominator in all these aspects is application, energy, bandwidth, and storage awareness, while the specific application contexts will have their respective signature requirements on energy capacity, processing capability, ambient environmental constraints, and quality of service (QoS)/quality of user experience (QoE) requirements.

First three subsections below outline the unique features of these user applications. In all cases, there is a pressing need for energy sustainability besides efficiency, because of continued miniaturization and convenience of wireless connectivity. Along with the possible smart grid connectivity for energy supply at different stages of the network architecture, these systems also are expected to be aided by ambient energy harvesting sources. This dual energy supply solution also adds to the possibility of two-way energy flow in the IoE. The fourth subsection deals with managing the energy grid connectivity that will have energy conventional generation sources, such as thermal and hydroelectricity, as well as the modern solar energy generators. Here, power grid network stability and load adaptability are the key aspects of IoE.

## 11.2.1    SMART MONITORING OF CIVIC INFRASTRUCTURE AND AMENITIES

In the era of IoT, the use of sensor network for has found many potent applications towards remote monitoring, control, and actuation, as well as in various automation. Smart agriculture, air and water pollution localization, terror alert in smart city, hazardous storage monitoring, home automation, object tracking, and smart healthcare are to name a few civilian applications [1]. Similar applications are also possible in military applications, such as monitoring border areas. A pictorial representation of the emerging IoT applications which require heterogeneous telemetric data collection and actuation is shown in Figure 11.1.

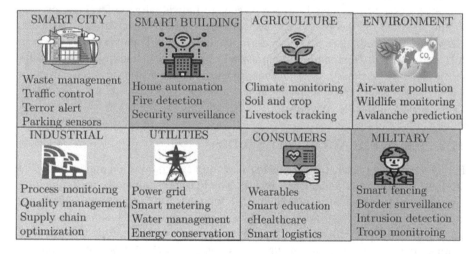

**Figure 11.1**    Emerging IoT applications contributing to IoE

With cost of smart devices being reduced substantially over the years, the technology becoming affordable to support these devices; thus more and more usage of these devices is becoming a common practice. To enable sustained usage, the basic need is to have many smart devices deployed for specific applications and in specific deployment environments. These devices are expected to be capable of autonomously performing the assigned tasks and communicate the required information, which can be either sensed/telemetric data or control data, to the required destinations for appropriate control actions as well as storage for future purpose. Some field IoT sensing applications, such as in smart cities, pollution monitoring, and agricultural automation, are characterized by high energy consumption, node lifetime criticality, and energy sustainability issues. A survey in [79] summarized the state-of-the-art in IoT area and sensing nodes. Usage of a low power sensing co-processor to enable continuous sensing for the smart phones was proposed in [67]. The applications in [53] and [31] also used low-power co-processor for energy efficiency of continuous sensing applications. The authors in [110] proposed a framework of energy-efficient mobile sensing system pertaining only to activity recognition applications. A framework for defining context sensing rules in mobile applications was proposed in [62] for energy saving by switching off unnecessary sensors. A measurement study on energy consumption of various components of a Smartphone was conducted in [11]. Collaboration across mobile devices of the users traveling along a fixed path was suggested in [80], to arrive at minimum energy sensing schedule. In a bid to save energy, the work in [74] proposed to exploit temporal correlation in the individual processes of interest in order to decide optimum duty-cycling of the sensor hub.

Collaboration among same type of spatially distributed sensors to achieve fine granularity or save energy is reported in prior art. For example, inter-sensor collaboration for tracking applications with objective of improving sensing quality with information-theoretic measures is discussed in [121]. Indoor monitoring of air pollutant (CO) with adaptive sampling rate of gas sensor in collaboration with passive infrared sensor was reported in [42]. The approach in [3] used less energy expensive sensors like accelerometer more often instead of more energy expensive GPS for location sensing. Environmental research community attempted to establish correlation across meteorological variables and pollutants in order to predict pollutant's concentration from other environmental parameters like temperature and humidity [100, 108, 119].

Monitoring and actuation data handling is intrinsically related to the network operation efficiency and convenience of uninterrupted network operation. To this end, automated energy replenishment of IoT devices to achieve energy sustainability for uninterrupted network operation is of paramount importance [46, 97]. To ensure long lasting autonomous operations of the field nodes, in addition to optimum sensor data sampling, it is critical to optimally awake the RF communication module at the field nodes, instead of relying on duty-cycling-based data delivery [64]. To this end, low-cost, low power wake-up radio and RFID based data collection techniques are of interest [87]. Although research on wake-up radio technology has been explored for quite some time [13, 18], cost and efficiency are still issues of relevant interest. Sensing, communication, and automation for uninterrupted operation of the IoT devices are expected to be aided by mobile energy supply and communication agents, such as ground vehicles [8, 78, 114], or UAVs [93, 98, 99, 118].

## 11.2.2  SMART WIRELESS SERVICES

In conventional wireless services, such as cellular wireless telephony, the primary objectives are coverage and QoS (e.g. data rate) support to the mobile users. User equipment (UE) energy optimization has been classically addressed by deciding on coding and transmission schemes that are light-weight on the downlink receiver side [68]. There also have been significant works on joint application, channel, and UE constraints and preferences aware energy optimization of the broadcast receivers [81, 82, 84]. Coverage load balancing using heterogeneous wireless technologies, such as cellular wireless (Long-term evolution (LTE)) and Wi-Fi have also been studied in recent literature [83, 85, 86].

Energy optimization at the base stations (BSs) have also drawn research attention [70, 72]. Inter-BS collaboration have also been addressed for energy saving and revenue sharing [63]. Stand-alone solar-operated BS energy optimization and CAPEX-OPEX balancing has also been studied recently. Solar dimensioning for an off grid, standalone base station has been presented in [92]. It considers powering a base station with solar power instead of the traditional power from the power grid. The paper gives a lower bound for the number of batteries and solar panel dimension required to operate a base station only considering the solar power. This work is further extended in [97], by finding the accurate solar dimensions with minimum cost requirement as well as reducing the complexity of the algorithm. It deals with finding the probability distribution of the left-over energy and computes the battery and panel size through it.

Recent attention on solar-powered and power grid connected BS marks an important highlight on the role of BSs or access points on IoE [54], [12]. A pictorial representation of multi-powered wireless services is shown in Figure 11.2. The objective has been to develop self-sustainable, cost effective and scalable communication strategies for the next generation (5G and beyond) wireless networks, with focus on resource allocation, QoS/QoE guarantee of the users, and cost profitability of the users as well as the service providers. The problem is motivated by the advent of IoT use cases, resulting in an increase in the number of network devices, ad hoc networks and the frequency of sporadic data transmission. Not only has it led to an increased congestion of the system bandwidth and interference but has also resulted in increasing contention for using the same channel. Alongside data, the net power consumption of the system is also increasing due to the presence of ad hoc nodes and network infrastructure to facilitate better network capacity. These devices need energy not only to communicate among themselves but also to self-sustain in the network, failing which the network performance may get hampered. Proliferation of IoT has also resulted in dynamically-varying, uneven user density in a coverage area. At a given time, while the mobile user density in some areas may be high, some other areas are sparsely populated. This imbalance results in congestion of some base stations while leaving some unused. Thus it is imperative to design communication network and resource allocation model according to the use case being presented. This requires to take into consideration various factors including heterogeneity of device traffic, nonuniform user densities, sporadic nature of data transmission by the devices, unreliability of the wireless channel, and randomness of ambient energy availability.

The systems aspects of importance for judging the quality and utility of network are the following. Dynamic prioritization of user traffic is required to guarantee QoS/QoE while aiming for resource optimization. Heterogeneity of user traffic and its dynamic nature of context prioritization is of interest for effective and efficient

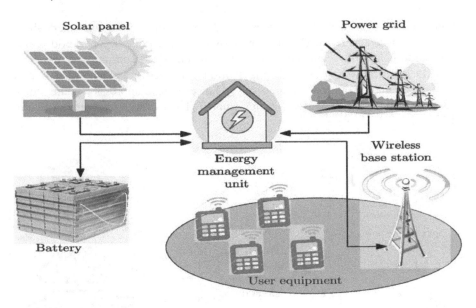

**Figure 11.2** Multi-powered wireless services

resource utilization. Energy sustainability and green resource utilization is a core aspect, wherein the dependence on non-renewable energy from power grid is targeted to be minimized. On the other hand, two-way energy flow in the IoE context is of interest, where the wireless service stations as conventional energy users also participate as energy suppliers. Finally, cost optimization is a paramount issue for the service providers, wherein CAPEX, OPEX, and short-term/long-term user satisfaction are optimized. Judicious balance between energy usage and wireless services through network level cooperation also helps energy cost minimization.

## 11.2.3 ADVANCED POWER METERING

Sensing devices installed in an IoT infrastructure generate vast chunks of data which is forwarded to the main system e.g. a server hosted on a cloud platform through a gateway for real-time or non-real-time decision making and/or actuation. Smart electricity meter is such an example of IoT device which are now-a-days being installed in domestic and industrial premises in order to monitor the power consumption along with various services like billing, load forecasting, and dynamic pricing. A smart meter includes various sensors, namely line frequency and phase currents and voltages. Besides, the auxiliary parameters such as system temperature and time stamp are also collected, for studying the electricity consumption pattern and system health, which are sent to the control center. It may also control the operation of the devices at the customer premises through command signals. On a regular basis, the utility provider receives the electricity consumption data from the customers' various devices. It helps the electricity service provider in managing the electricity demand-response in an efficient manner. It also provides useful information to the consumers about their energy usage patterns. The integration of smart meters in the smart grid is also expected to help utility companies to identify unauthorized consumption and electricity thefts, thereby improving the distribution efficiency and quality of power. A block representation of near-real-time sensor data pruning/compression at the source and compression at the cloud server in a smart IoT network is presented in Figure 11.3.

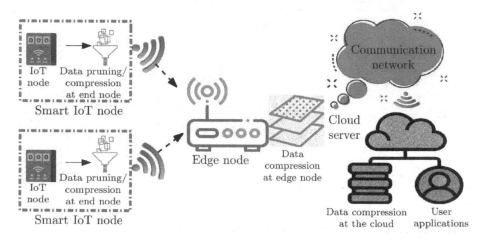

**Figure 11.3** Example of a Smart IoT network

Advanced Metering Infrastructure (AMI) has emerged as a powerful means for two-way communication between the end consumers and the utility service providers. Besides smart wireless monitoring, smart meter based fast monitoring of power consumption of appliances is another big data generating application. Naturally, the foremost concern has been how to reduce the data volume from communication as well as cloud storage viewpoints. Smart meter data patterns suggest that, it has a unique nature of data distribution and dynamics. Since the data captures the electricity consumption patterns of home or business appliances, it causes vulnerability to the users in terms of privacy and security concerns. Furthermore, the access network connectivity could be more ad hoc, such as over WiFi, or low-power Wide Area Network (e.g. LoRa-WAN), or Narrow-Band IoT links that are operated by multiple service providers. The smart meter data can be easily commercially exploited by third party. Therefore, dealing with smart meter data requires a different approach in order to securely and cost-effectively communicate. A study in [19] proposed compression of large datasets in smart grid based on singular value decomposition. Dictionary learning and sparse encoding was used in [112] to decompose the load profile into partial usage patterns and achieve compression of smart meter data. In [103], generalized extreme value characterization was proposed for feature based load data compression to identify load states and events. A resumable compression method based on differential and variable length encoding was studied in

[109]. Another study in [4] represents smart meter readings as Gaussian waveforms with minimum features and applies burrow-wheeler transform and entropy encoding for its compression. Linear Gaussian model-based load profiling was proposed in [90] to capture behavioral pattern of residential customers. Exponential, lognormal, generalized Pareto, beta, and gamma distributions are some other models considered in [90, 103] (and references therein) to characterize the smart meter data. A more recent and accurate technique of high frequency smart meter data characterization uses Gaussian mixture model (GMM) [105], which was shown to have a significantly better fit compared to the existing models.

More discussion on smart meter data compressibility and novel approaches to dealing with big data concerns caused by millions of smart meter installation as a part of smart city monitoring exercise are presented in Section 11.3.

### 11.2.4  SMART GRID MONITORING

Power grid connectivity classically serves the electrical power needs for household, office, and industrial energy demands. Conventional power generation sources are dedicated thermal and hydro power stations. Although the power generation stations have predictable rate of generation, the power consumption patterns are somewhat unpredictable. Therefore, even in the conventional power transmission and distribution systems, there are chances of generation versus consumption imbalance that may lead to instability of power grid. Stress on power grid causes frequent occurrences of failure events, raising questions on safe and reliable grid operation. Hence, for uninterrupted power supply there is a need for monitoring the disturbances in power line. The parameters of power grid health monitoring interest include phase voltages and currents, their phasor values, and the powerline frequency. Intuitively, imparting intelligence, automation, and control to the traditional power grid can enhance its load handling capabilities. An old system for power grid monitoring is by using supervisory control and data acquisition (SCADA). In the modern smart grid this is being replaced by a more network connectivity friendly monitoring modules, known as phase monitoring units (PMUs). PMUs periodically sample the state of power system and send their data to a remotely located phasor data concentrator (PDC) in a control center over a communication network [35]. An example of PMU placement across the smart grid network in a wide area monitoring and control system (WAMCS) is depicted in Figure 11.4.

As the modern day power grid is more and more connected to renewable energy sources, such as solar and wind energy, the power generation is no longer predictable because of dynamic stochastic nature of the renewable energy resources. Further, with the advent of massive IoT deployment in smart cities and increased demand for self-sustainable wireless services, commercial as well as domestic service points, which used to be conventionally solely the electricity consumers, are also stating to generate renewable energy. Energy demands of these isolated wireless stations as well as microgrids with renewable energy sources as well as their energy surpluses are sporadic. The surplus energy is fed back to the macro-grid [120]. With the distributed power generation into the power grid and sporadic demand from increased IoT service nodes, unpredictability of power generation and load further in-

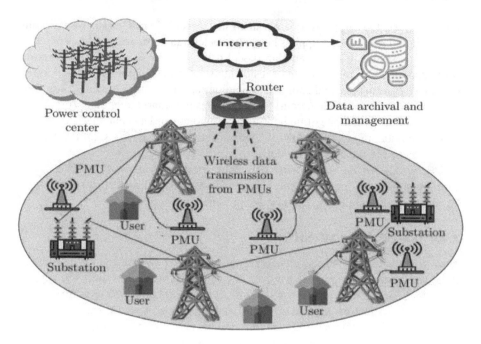

**Figure 11.4**   Basic elements of a wide area monitoring system

creases, which lead to more frequent perturbation on powerline frequency and phasor voltages and currents [91]. Another aspect of instability causing behavior in modern power grid is that, unlike the conventional power generators that are associated with mechanical inertia, the renewable energy sources are not associated with any inertia. As a result, when a microgrid chooses islanding from, or resynchronization into the power grid, perturbation is more severe. The resultant effect is that, the need to monitor the smart power grid health using PMUs has become more significant.

Clearly, data from the PMUs play a crucial role at the PDC in accurately determining the health of the power system [32]. Since the PMUs are mere IoT sensing devices that are not imparted with any decision-making capability on the power grid network, it is expected that the data collected at the PDC from multiple PMUs are analyzed on real-time basis for power generation and load distribution related decision making. Thus, reliable and rapid coordination between the PMUs and the PDC is essential to capture the power grid dynamics and provide real-time situational awareness and actionability, thereby ensuring good health of power grid. It is notable that, this data-level intelligence and functionality in smart grid is in addition to the automated relay and mechanical circuit protection features to locally arrest the power line faults. More specifically, using the PMU-PDC data network the power line fluctuations that are not traceable by these electro-mechanical devices are analyzed at a cluster or central level for a high-level intelligence making and power load balancing. A major aspect of data analysis for short-term and long-term decision making includes cyber-attacks in the form of node-level compromise, data spoofing, denial of service attacks, etc. The data in central repository or remote cloud is further used for long-term strategic decisions on the power grid and various data analytic purposes.

Currently there is no unique or preferred solution on WAMCS connectivity, network infrastructure, or architecture. The possible current-day alternatives for smart grid connectivity and information communication infrastructure include augmenting the high power electricity carrying powerline signals, subscribing to the cellular mobile/wireless network connectivity technologies, namely, Universal Mobile Telecommunications Service, Long-Term Evolution (LTE), LTE-Advanced, Wireless Local Area Networks, and Ethernet, and developing a dedicated network for smart grid signaling, e.g. optical fiber network connectivity. The network architecture depends on the type of networking technology chosen. It could be optical fiber ring or bus connectivity, or heterogeneous wireless technologies, or a loosely-connected hierarchical structure involving both of wireless and wireline infrastructure – as in the Internet. Each of the possibilities have their respective advantages and shortcomings. While a dedicated network architecture ensures resource guaranty, reliability, low latency, and more control over data compromise, cost and scalability aspects suggest to use the already-existing information communication network infrastructure, primarily because the data from the individual PMUs is perceived to be of low-rate telemetric nature, requiring small bandwidth. When using the existing communication network infrastructure for PMU-PDC connectivity, shared network resources introduce uncertainty on timely delivery and actionability to the remote cloud. Here, some hierarchical structure of connectivity involving local PDC and remote PDC is suggested, with distributed intelligence for timely action from the local PDC [61].

On one hand, the emergence of IoE is expected to generate huge amount of data on human and machine activities, offering newer insights on remote controllability and automation of various processes and tasks. *Big data* analytics via Artificial Intelligence and Machine Learning (AIML) using powerful machines is already playing a huge role in extracting the optimal solutions to very complex tasks [71]. At the same time, light-weight version of ML (LWML) for resource-constrained miniature IoE devices is emerging as a popular data-driven decision optimization. On the other hand, to ensure that wireless services deployment and operation in the context of smart cities would be scalable and *energy sustainable*, it is important to have a stable two-way energy flow between the power grid and the wireless stations connected with renewable energy source. The optimization tasks in this objective involve multidisciplinary efforts, namely, node/network-level energy efficiency optimization, data-driven learning-based solutions, device-level embedded implementation, novel solutions to on-demand node-level energy replenishment, smart grid communication network architecture design with distributed intelligence, and joint traffic and energy load balancing [20, 21]. Some of these aspects are discussed in the following two sections.

## 11.3  ROLES OF BIG DATA AND CONTEXT-SPECIFIC LEARNING IN FUTURE IOE

In future IoE it is expected that machine-to-machine and human-to-machine communications will significantly dominate the conventional human-to-human interaction via communication networks. This also implies that, much data needs to be communicated and processed by the machines for close-to-human ability of decision making. Thus, big data communication, storage, and processing/analytics will play major role [88]. The following two subsections highlight context-specific roles, challenges, and the new paradigm of data-driven optimization strategies. While there are multitude of evolving gadgets in future IoE, a few example IoT applications are highlighted that are expected to have large-scale impact in future smart cities.

### 11.3.1  ROLES AND CHALLENGES OF BIG DATA

To illustrate the roles and challenges of big data, we first consider two typical modern IoT applications. In smart grid data communication from PMUs to PDC, while frequent sampling and real-time data reporting facilitate in timely control actions for preserving grid stability, high volume of data is a challenge for communication bandwidth and storage needs [43, 51]. For instance, a study in [66] has noted that data transmitted from 100 PMUs at 30 samples/s to a PDC is over 50 GB data per day. A related big data handling application has been smart meter based fast monitoring of power consumption of appliances. Millions of smart meters that are being installed across the world for commercial and domestic power consumption monitoring purposes generate large volume of data. As noted in [6]. The accumulated volume is expected to be up to the order of hundreds of terabytes per year in near future. Considering practical examples, it was recently shown in [106, 113] that, although at the individual level a smart meter may generate about 5 kB of data in every 15 minutes, collective volume from a hundred million of such devices can be as high as 2920 TB in a year. Usually these IoT devices sample at high rate and report the whole data to the cloud. In this process large amount of energy is consumed at the IoT node for transmission of the data, large bandwidth is used for transmission over the Internet, and also storage requirement to save this enormous data is very high. However, resource-efficient communication and storage of this vast data remain a challenging task. From the perspectives of the Internet service providers' bandwidth consumption as well as storage cost, loading the network unnecessarily is undesirable [102, 105, 106, 113].

A similar case of data explosion arises from wireless IoT sensor networks from multitudes of applications, namely, environmental monitoring, industrial process monitoring, smart home, smart agriculture, smart wearables, remote healthcare, city and remote area surveillance, etc. Typically, a node has multiple sensing parameters, the sensors (such as the electro-chemical reaction or thermal heating based pollution sensors) could be individually energy hungry [36], and there could be correlation among data collected by the different sensing elements [37]. Depending on the application context, data sampling rate could be high, on the order of once in every few seconds. Further, Often, the nodes are densely-deployed, to ensure full area coverage and sufficient redundancy of data to account for the node failure as well as communication-related data loss. All these factors lead to the possibility of very high volume of data to be transmitted over a field node's wireless interface (causing energy drainage), to be communicated over the network (causing high bandwidth consumption) and stored at a cloud repository (requiring large storage space and the associated cost) [23]. Since in most smart environment applications the IoT nodes are wirelessly connected to the Internet, dealing with high volume of data at the node level also raises the concern of node-level energy sustainability and uninterrupted network operation [38, 74].

A related issue is dealing with large volume of data generated at the end nodes and making appropriate intelligence and timely actuation capability [2, 7, 49, 52]. On one hand, carrying all data to the central cloud and relying on data analytics by powerful computation may be computation-resource-wise efficient, but may not be acceptable in some time-critical applications, such as tele-surgery, smart grid monitoring, etc., due to high network latency. Moreover, carrying all field sensing traffic without applying local intelligence on their usefulness may prove communication and energy resource-wise inefficient. On the other hand, applying intelligence at the sensing traffic generation process or pruning as an after-sensing exercise may be of interest from communication bandwidth and transmission energy efficiency perspectives. Additionally, having some intelligence at the node-level may be helpful in time-critical decision making. However, the processing overhead could be a bottleneck at the node with respect to its processing capability (equivalently, its cost) as well as processing energy requirement. With both the alternatives, the factor of data vulnerability and security at the processor level as well as at the network routing/switching level are important factors in deciding an option. Thus, an application context specific IoT node architecture design as well as network architecture design in deciding on distributing the intelligence is of interest [2, 7].

## 11.3.2   NODE- AND NETWORK-LEVEL DATA-DRIVEN OPTIMIZATION

As alluded in the previous subsection, dealing with big data in IoE requires proper understanding about the application context at hand and the constraints that are in place. Accordingly, the optimizations and solutions to the problems are expected to be different. It may be noted here that, commercially-available IoT hardware devices, namely, sensor hubs, smart-meters, phasor measurement units (PMUs), do not have the provision for data pruning at the field node. Likewise, the distribution of intelligence in the network for resource-optimized data handling is yet to be matured.

We highlight two example application cases: one is on real-time application (smart grid monitoring); the other is on not-so-time-critical application (smart metering, wireless sensor networks), to discuss how the solution approaches evolve. The focus is given on intelligent data pruning at the end node for transmission energy, communication bandwidth, and cloud storage space optimization.

**Smart grid monitoring:** On high-rate sampled PMU data handling for real-time smart grid monitoring, a few studies have explored stability prediction based on reduced PMU data. In [14], autoregressive modelling of PMU data sequence is studied to identify stress signs from correlation between consecutive samples. Short-term prediction using state-space approach and basis function was studied in [22] to spot measurement errors. Dimensionality reduction of PMU data using linear principal component analysis were studied in [30, 115]. Discrete cosine transform [77], [69], compressive sampling [15], wavelet packet decomposition [26, 47], preprocessing and lossless encoding [101] (and the references therein) operate offline and also in real-time for data storage. A few related studies in context of cloud storage are optimal sensor selection [9], K-means clustering [65], use of support vector machine for stored data analysis [89]. It is notable that, other than [15], all prior approaches investigated data reduction at the PDC. However, although the objective in [15] has been communication bandwidth reduction, it does not deal with nonstationary nature of PMU data. Thus, in absence of continuous learning and adaptation, quality of compression and hence the quality of power system health monitoring is expected to degrade over time.

In a recent study it was demonstrated that the PMU data on smart grid health is truly non-stationary in nature, which implies that classical theory of stochastic processes will lead to non-optimal solutions [104]. Hence, dynamic learning-based analysis for data pruning at the PMU is necessary. A learning-based framework based on $\varepsilon$–support vector regression was proposed to dynamically prune the PMU data before transmission. In the proposed approach, learning parameters are dynamically updated as necessary to keep the error in data pruning within a specified error limit. Follow-up work in [107] further incorporated communication channel related uncertainties in deciding the optimal channel overhead required to be incorporated at the IoT data source node (PMU) to ensure that the pruned data content is successfully delivered. Besides a stochastic analysis based proposition, a learning-based channel overhead computation was also proposed to show the benefit on accurate overhead prediction – albeit at some cost of processing overhead.

In the current state of the art, the PMU data is still used for aiding human intelligence in smart grid traffic load balancing decision making. The mechanical relays locally address the major faults in the transmission/distribution lines. To this end, some critical aspects of interest that require further studies are (i) understanding vulnerabilities in the smart grid network based on the PMU data, (ii) analysis of smart grid health based on multiple neighboring PMU data by exploiting the possible correlation, (iii) investigating on distributed intelligence placement for automated actionability on PMU data for actuation of energy traffic redistribution on real-time basis, (iv) studying the network routing and switching related nonidealities and their impact on PMU-PDC coordination. Although the above-mentioned directions of future research pertain to smart grid monitoring and control applications, the propositions would apply broadly to the other time-critical (ultra-low latency aware) and reliable IoT applications.

**Smart metering:** A semi-real-time or delay-tolerant IoT application is smart metering, on which there have been recent efforts on data reduction without compromising on quality of information. Since smart metering is a less time-critical application, pruning technique of high-rate sampled data at the source is compression-based. The conventional signal processing algorithms for data pruning are entropy coding [4], singular value decomposition [19], load features based compression [103], and dictionary learning and sparse encoding [112], that exploit the temporal and spatial attributes of the data streams from different sources collected at an aggregation point. These techniques do not address the issue of single time-series data reduction at the source, and thus are less useful in near-real time applications. A lossy compression method [24] produces piece-wise approximation of original data to control the smart meter data volume at the cost of accuracy. To address this, performance of four loss-less compression algorithms (adaptive trimmed Huffman, adaptive Markov chain Huffman, tiny Lempel Ziv Markov chain, and Lempel Ziv Markov chain Huffman) were investigated in [73, 116]. A resumable compression method [109] based on differential coding also proposes compression of data sampled at 1 second interval for household level and 3 seconds interval for appliance level at the smart meter. It was observed that, although high granularity (on the order of seconds) is critical to attain substantial compressive gains, compression of high resolution household level data at the smart meter remains challenging owing to rapidly fluctuating load patterns. To this end, an adaptive data reduction scheme using compressive sampling was devised in [105], to operate at the source which achieves significant bandwidth saving in data transmission to the nearest collection center without any appreciable loss of information. Unlike other data compression algorithms, this compressive sampling based data reduction exploits the rapidly fluctuating nature of high resolution sensor data. It was observed that, although the data appears incoherent in time domain, it can actually be concisely represented in a sparsifying basis. Thus, adaptively choosing the sparsity over optimum batch size before data transmission can be utilized for substantial reduction in data volume.

Data pruning in a more generalized scenario is of interest wherein the IoT node has multiple sensing elements which generate high-rate multiple time series of near-real-time data. Intuitively, extending the concept in [105] over each individual time-series data would help data reduction over the individual streams. Anticipating possible correlation among different measured variables, in a recent study [75] cross-correlation of the streams along with the temporal correlation on the individual time-series data were exploited to reduce the communication bandwidth as well as cloud storage requirements. For capturing distribution of the original data and comparing the quality of compression, joint distribution of multivariate data was obtained using Multivariate Normal Autoregressive Integrated Moving Average model. A multivariate data compression scheme was devised which exploits cross-correlation among different variables measured at an IoT node to reduce the transmitted data dimension. The scheme follows a two-step adaptive compression. In the first step dimensionality of data is reduced by applying Principal Component Analysis. In the second step, each of those selected components is further compressed temporally using Compressive Sampling on the reduced number of streams in the transformed domain. In principle, this data reduction technique is a powerful approach in any multivariate IoT sensing applications that is expected to yield significant gain in transmission energy, bandwidth saving, as well as storage space requirements. The resultant pruned data from the field node is interpolated at the destination using a counterpart two-step decompression scheme. This process is expected to aid in mitigating the big data footprint in massive IoT communication scenario.

An important aspect is incorporation of the dynamic data compression technique at the source, thereby developing advanced smart meters, in a cost-effective manner so as to ensure commercial viability of reduced data generation from such IoT nodes. While the compression principle remains the same, there may be vendor-specific different techniques of achieving the data pruning. Therefore, compatibility standard in terms of interfacing with the Internet, such as light-weight machine-to-machine (LW-M2M) interface, may be required to ensure equipment modularity.

**Wireless IoT based smart environments:** While data handling in the above two applications (smart grid monitoring and smart metering) are focused on communication bandwidth saving and cloud storage saving, in other smart city IoT applications the data reduction is motivated by energy saving at the IoT nodes, especially because of their wireless connectivity and deployment in remote/hazardous locations [48]. The other unique features and possible novel solutions include the following: (i) Data generated may be mostly telemetric, i.e. delay-tolerant. However, there may be applications, such as remote health, where the data priority may have to be dynamically adjusted depending on the instantaneous nature of data. (ii) Since sensing or sampling itself could be energy-hungry, such as in pollution monitoring, high-rate sampling and then data pruning before transmission may not be energy-efficient. Therefore, node-level [74] and network-level [37, 40] collaborative sensing techniques may be required to prune the data at the generation process itself. Depending on whether an IoT node has single parameter or multiple parameters of data to sense and its current energy availability, newer supervised (single/multi-parameter support vector machine) or unsupervised learning (reinforcement learning, context-aware bandit algorithms) techniques may be employed. Low-cost wakeup features for sensing as well as communication may be further explored to minimize energy consumption [45].

Since the question of node-level energy sustainability as well as cost-affordability are of key importance, appropriate decision on where to incorporate the learning and processing intelligence is critical. This is clearly application and ownership context-specific. Distributing intelligence appropriately at the end (fog) nodes or at the immediate collection point (edge node) or at a higher level of network connectivity hierarchy is required. An evolving direction of research in this context is mobile edge computing [2, 7].

## 11.4    ROLE AND CHALLENGES OF SMART GRID IN IOE ENERGY SUSTAINABILITY

### 11.4.1    ENERGY SUSTAINABILITY

While big data handling and edge computing address the resource efficiency (bandwidth, storage, energy), energy sustainability still requires major attention in order to make IoE a mass reality. Without convenient recharging capability and low CAPEX and OPEX, large-scale adoption and deployment of IoT technologies may not happen [96]. Even for the IoT equipment that may be connected to the power grid supply, drawing power from ever-depleting fossil resources (coal, oil) are discouraged. Thus, powering devices and systems from ambient (green) resources, such as the energy from solar irradiance, wind, pressure, heat, water current, and ambient RF, has got significant traction in communication systems and networks research community [10, 12, 25, 41, 46, 54, 60, 97]. Also, in situations where ambient energy resources may not be sufficient at times, convenient means of recharging of the IoT devices are of interest. Wireless energy transfer, in the form electromagnetic inductive coupling and RF radiation, are some competitive/complementary techniques that are being pursued [16, 34, 44, 45, 50, 55, 56, 57, 93, 98, 99, 111, 114].

**Green energy:** On the green energy sustainability solution solely from ambient sources, one aspect is to design the energy harvesting antenna and storage battery/super-capacitor dimensioning, as a CAPEX estimate [94, 95]. This, combined with the stochastic analysis of ambient energy availability, analysis of traffic variability and heterogeneity distribution, energy usage optimization (discussed in Section 11.3.2), and the user QoS/QoE support, decide the OPEX, and combined CAPEX-OPEX optimization is aimed [46, 97]. Ambient RF energy harvesting is another proposition, primarily in the mobile wireless research community, wherein the ambient RF energy is cognitively harvested by the same receiver antenna when either the node is not communicating but there is sufficient ambient RF energy available or energy harvesting is estimated to be more cost-effective than communication at that instant [60]. The RF energy could be available in-band (cellular or WiFi, over which the node also communicates when required), or it could be out of band, e.g. over digital television band – in which case additional tuning capability at the energy harvesting receiving node is required. Harvesting capability, sensing activity, and communication optimizations in this direction depend on the nature of traffic/data and QoS to be served.

**Wireless recharging:** Wireless energy transfer is a complementary solution to recharge the wireless IoT nodes on-demand. Wherever ambient energy sources are insufficient and power grid connectivity is unavailable, robot-augmented wireless energy transfer is being explored for sustainable IoT network coverage and communication [16]. On one hand, beyond the commercial state-of-the-art (www.powercast.co) there has been systems-level studies to increase the RF energy transfer efficiency via novel multihop energy routing to the field IoT node [44, 46, 55, 56]. On the other hand, there are newer wireless RF charging strategies using mobile vehicles or robots or unmanned aerial vehicle (UAV) that are being explored, targeting various sensing applications, such as pollution monitoring in hazardous locations, smart agriculture, surveillance [16, 98]. Besides poor RF energy harvesting sensitivity and energy rectification efficiency, the other unique aspects of research interest here are: (i) mobile RF charging robot's path planning for charging multiple field nodes with stochastic charging as well as discharging pattern, and (ii) nonidealities of system and charging environments.

**Smart grid connectivity:** The third aspect of energy sustainability with possible green energy resources features an evolving scenario where the nodes have the ambient energy harvesting capability along with their power grid connectivity [54], [12]. A practical example is cellular BSs in semi-urban or rural areas in developing countries, where power from the grid connectivity is not always guaranteed. Even in smart city context, in a bid to green energy sustainability, the power-line connected wireless access points could be equipped with ambient energy harvesting capability. Having a dual power connectivity (e.g. solar energy and power grid) helps in design and dimensioning of solar panel and storage batteries without looking to factor the worst-case scenarios of traffic intensity and sustained nonavailability of solar energy. While solar (ambient) energy availability is stochastic, in most mobile communication as well as IoT sensing applications the traffic pattern that a node needs to handle is also dynamically varying. Thus, in the worse-case scenario, green energy resource dimensioning may be insufficient, requiring the node to draw power from the power grid. Conversely, in the best-case scenario the node may have sufficient surplus energy which may be transferred to the power grid. In a networked set of BSs or wireless access points, these situations of deficient or surplus power can be dealt with further efficiently, for cost optimization of the network service provider.

A fallout of this two-way power flow between a wireless (IoT edge) node and power grid is that the edge nodes also act as energy supply points – though with sporadic nature, aiding to the power grid's energy flexibility. In the next section we will discuss the additional challenges in power grid that are associated with the two-way energy flow.

## 11.4.2   STABILITY AND CONTROLLABILITY OF POWER GRID

**Challenges and mitigation approaches:** Conventional, old SCADA system that is being replaced by the PMUs in smart grid aim at timely reporting of impending instability and consequent blackout in power grid. This is caused by extreme imbalance between power load demand and supply. With the PMU-aided smart grid, real-time prediction and mitigation of blackout are possible through robust and secure extraction of stability information from the health parameters collected at the PMUs. One challenge is that, the power grid health anomaly data collected at a PMU does not necessarily imply it is caused the nearby disturbances. To make a better sense on the actual health status and the causes, it is important that the data from multiple PMUs are jointly analyzed. To explore joint data analysis, the data from multiple PMUs need to be studied at a collection point. To enable timely intervention for grid failure mitigation, appropriate clustering of PMUs for distributed data processing is required that can minimize communication and processing delay [59]. Adaptive corrective controller design for real-time response are necessary for automated response to the predicted instability.

As noted in Section 11.2.4, while the PMUs capture the load fluctuation through line frequency variation and other phasor data, the major faults occurring in the grid are arrested by the mechanical relays. However, if the quantum of perturbation at some points are below the circuit breaking sensitivity of the mechanical relays, such perturbations continue to propagate beyond the fault location region. It is important here to analyze how multiple neighboring PMUs capture such events. It is also important to investigate how fast and along which trajectory the perturbations propagate across the network. The information on disturbance propagation, if accurately extracted, can be used for isolating the faults, thereby avoiding costly cascaded failure of the power grid.

In order to controllably handle big data issue, the works in [104, 107] have suggested to introduce data pruning. However, unless the network vulnerabilities are sufficiently addressed, data pruning may invite insufficient statistics of grid health data to reach at the PDC, which may trigger false actions. Security breaches at the node level or at the network may also wrongly let the systems manager take a wrong decision on power grid health. Therefore, data pruning technology needs to be accompanied by additional capabilities of robustness at the system level data transmission. Further, contemporary smart grids still do not have the automation feature. The data is merely used for human intelligence on load distribution through manual intervention. These are some of the aspects that need attention for fail-safe smart grid automation.

**Modern-day power grid challenges:** While conventional power grid has its own issues of stability and predictable controllability, the introduction of two-way energy connectivity in smart city context adds to its further challenge [117]. The additional power supply from microgrids act as distributed power generators, which is a major step forward to reducing the carbon footprint while directly/indirectly serving massive IoT connectivity. A challenge is also posed by intermittent islanding and resynchronization of the microgrids [33]. Since surplus power from the microgrids are from ambient sources, there is no inertia involved with power generation. Likewise, depending on the microgrid's power requirement when it decides to isolate itself (through islanding) from the power grid, sudden change occurs in power supply/demand from the microgrid. Combinedly, random nature of supply-demand causes load fluctuation in the power grid, which is manifested in its change in line frequency. A high magnitude of change of load leads to a higher change of frequency, which may lead to instability in the power grid. To address the stability issue, a possible mitigation technique is to introduce a set of protocols at the islanding and resynchronization stages [76]. With the renewable energy which is already a DC signal and most of the IoT appliances operating with DC power, a DC power grid and its distributed control has evolved [58]. The nature of data pruning, associated with stochastic loss and delay behavior of smart grid data routing also add to the stability issue. Data compromise at the routing stage or spoofing the data transmitted over public network are some additional challenges that are of contemporary research interest.

## 11.5   CONCLUDING REMARKS

Through this chapter a paradigm shift associated with smart environments is highlighted. It has been demonstrated how the challenges of big data and energy sustainability are addressed via application-specific unique approaches. For example, the data-driven and network solutions for dealing with real-time (low-latency) traffic has been very different compared to those associated with not-so-time-critical but loss-intolerant traffic. Since each IoE application has the corresponding specialized features and device-level constraints, it has been shown that the network intelligence solutions need to be appropriately designed. Beyond energy/bandwidth/storage efficient optimization techniques, energy sustainability and green communication techniques have been presented. The research state-of-the-art and open issues on mitigating big data challenges and energy sustainability have been highlighted. Further, the role of smart grid on uninterrupted IoE operation, the benefits of distributed energy generation through IoE, and the newer challenges of power grid stability/controllability have also been discussed.

## ACKNOWLEDGMENT

This work was supported in part by the Department of Science and Technology (DST), International Bilateral Cooperation Division, under Grant INT/UK/P-153/2017, and in part by the Science and Engineering Research Board, DST, under Grant CRG/2019/002293.

## REFERENCES

1. Document 3GPP TS 22.368 V13.2.0 (2016). Service requirements for machine-type communications (MTC).
2. Abdellatif, A. A., Mohamed, A., Chiasserini, C.-F., Tlili, M., & Erbad, A. (2019). Edge computing for smart health: context- aware approaches, opportunities, and challenges. *IEEE Networks, 33(3)*, 196–203.
3. Abdesslem, F. B., Phillips, A., & Henderson, T. (2009). Less is more: Energy-efficient mobile sensing with senseless. In *Proc. ACM MobiHeld*, New York, NY, USA, 61–62.
4. Abuadbba, A., Khalil, I., & Yu, X. (2017). Gaussian approximation based lossless compression of smart meter readings. *IEEE Transactions on Smart Grid, 9(5)*, 5047–5056.
5. Ahmed, E. & Rehmani, M. H. (2017). Mobile edge computing: Opportunities, solutions, and challenges. *Elsevier Future Gen. Comput. Sys., 70*, 59–63.
6. Aiello, M., & Pagani, G. A. (2014). The smart grid's data generating potentials. In *Proc. Federated Conf. Comput. Sci. Inf. Syst.*, Warsaw, Poland, 9–16.
7. Avino, G., Bande, P., Frangoudis, P. A., Vitale, C., Casetti, C., Chiasserini, C.-F., Gebru, K., Ksentini, A., & Zennaro, G. (2019). A MEC-based extended virtual sensing for automotive services. *IEEE Trans. Netw. Service Manag, 16(4)*, 1450–1463.
8. Baroudi, U. (2017). Robot-assisted maintenance of wireless sensor networks using wireless energy transfer. *IEEE Sensors J., 17(14)*, 4661–4671.
9. Bijarbooneh, F., Du, W., Ngai, E., Fu, X., & Liu, J. (2016). Cloud-assisted data fusion and sensor selection for internet of things. *IEEE Internet of Things J., 3(3)*, 257–268.
10. Cammarano, A., Petrioli, C., & Spenza, D. (2016). Online energy harvesting prediction in environmentally powered wireless sensor networks. *IEEE Sensors J., 16(17)*, 6793–6804.
11. Carroll, A. & Heiser, G. (2010). An analysis of power consumption in a smartphone. In *Proc. USENIX Annual Technical Conf.*, Berkeley, CA, USA.

12. Chamola, V., Sikdar, B., & Krishnamachari, B. (2017). Delay aware resource management for grid energy savings in green cellular base stations with hybrid power supplies. *IEEE Trans. Commun., 65(3)*, 1092–1104.
13. Chen, L., Cool, S., Ba, H., Heinzelman, W., Demirkol, I., Muncuk, U., Chowdhury, K., & Basagni, S. (2013). Range extension of passive wake-up radio systems through energy harvesting. In *Proc. IEEE ICC*, Budapest, Hungary, 1549–1554.
14. Cotilla-Sanchez, E., Hines, P. D. H., & Danforth, C. M. (2012). Predicting critical transitions from time series synchrophasor data. *IEEE Trans. Smart Grid, 3(4)*, 1832–1840.
15. Das S. & Sidhu, T. S. (2014). Application of compressive sampling in synchrophasor data communication in WAMS. *IEEE Trans. Ind. Informat., 10(1)*, 450–460.
16. De, S. & Singhal, R. (2012). Toward uninterrupted operation of wireless sensor networks. *IEEE Computer Mag., 45(9)*, 24–30.
17. Jindal, A., Kumar, N., & Singh, M. (2020). Internet of energy-based demand response management scheme for smart homes and PHEVs using SVM. Future Generation Computer Systems, 108, 1058–1068.
18. De Donno, D., Catarinucci, L. & Tarricone, L. (2014). Ultralong-range RFID-based wake-up radios for wireless sensor networks. *IEEE Sensors J., 14(11)*, 4016–4017.
19. de Souza, J. C. S., Assis, T. M. L., & Pal, B. C. (2017). Data compression in smart distribution systems via singular value decomposition. *IEEE Trans. Smart Grid, 8(1)*, 275–284.
20. Aujla, G.S., Jindal, A. and Kumar, N., 2018. EVaaS: Electric vehicle-as-a-service for energy trading in SDN-enabled smart transportation system. *Comput. Netw.,* 143, 247–262.
21. Jindal, A., Aujla, G. S., Kumar, N., & Misra, S. (2018, May). Sustainable smart energy cyber-physical system: Can electric vehicles suffice its needs?. In *2018 IEEE International Conference on Communications Workshops (ICC Workshops)* (pp. 1–6). IEEE.
22. Dong, J., Ma, X., Djouadi, S. M., Li, H., & Liu, Y. (2014). Frequency prediction of power systems in FNET based on state-space approach and uncertain basis functions. *IEEE Trans. Power Syst., 29(6)*, 2602–2612.
23. Ejaz, W., Anpalagan, A., Imran, M. A., Jo, M., Naeem, M., Qaisar, S. B., & Wang, W. (2016). Internet of things (IoT) in 5G wireless communications. *IEEE Access*, 4, 10310–10314.
24. Eichinger, F., Efros, P., Karnouskos, S., & Bohm, K. (2015). A time-series compression technique and its application to the smart grid. *VLDB J., 24(2)*, 193–218.
25. Escolar, S., Chessa, S., & Carretero, J. (2014). Energy management in solar cells powered wireless sensor networks for quality of service optimization. *Personal and Ubiquitous Computing, 18(2)*, 449–464.
26. Gadde, P. H., Biswal, M., Brahma, S., & Cao, H. (2016). Efficient compression of PMU data in WAMS. *IEEE Trans. Smart Grid, 7(5)*, 2406–2413.
27. Gajduk, A. Todorovski, M., & Kocarev, L. (2014). Stability of power grids: An overview. *Eur. Phy. J. Spl. Topics, 223*, 2387–2409.
28. Kumar, N., Aujla, G. S., Das, A. K., & Conti, M. (2019). ECCAuth: A secure authentication protocol for demand response management in a smart grid system. *IEEE Trans. on Industr. Inform, 15(12)*, 6572–6582.
29. Kaur, D., Aujla, G. S., Kumar, N., Zomaya, A. Y., Perera, C., & Ranjan, R. (2018). Tensor-based big data management scheme for dimensionality reduction problem in smart grid systems: SDN perspective. *IEEE Trans. Knowle. Data Eng., 30(10)*, 1985–1998.

30. Ge, Y., Flueck, A. J., Kim, D. K., Ahn, J. B., Lee, J. D., & Kwon, D. Y. (2015). Power system real-time event detection and associated data archival reduction based on synchrophasors. *IEEE Trans. Smart Grid, 6(4)*, 2088–2097.

31. Georgiev, P., Lane, N. D., Rachuri, K. K., & Mascolo, C. (2014). DSP. Ear: Leveraging co-processor support for continuous audio sensing on smartphones. In *Proc. ACM SenSys*, New York, NY, USA, 295–309.

32. Ghosh, D., Ghose, T. & Mohanta, D. K. (2013). Communication feasibility analysis for smart grid with phasor measurement units. *IEEE Trans. Ind. Informat., 9(3)*, 1486–1496.

33. Grewal, D. S., Duggal K., & Sood, V. K. (2011). Impact of islanding and resynchronization on distribution systems. In *Prof. IEEE Electrical Power and Energy Conf.*, Winnipeg, Canada, 6–10.

34. Griffin, B. & Detweiler, C. (2012). Resonant wireless power transfer to ground sensors from a UAV. In *Proc. IEEE Int. Conf. Robot. Autom.*, Saint Paul, MN, USA, 2660–2665.

35. Gungor, V. C., Sahin, D., Kocak, T., Ergut, S., Buccella, C., Cecati, C., & Hancke, G. P. (2011). Smart grid technologies: Communication technologies and standards. *IEEE Trans. Ind. Informat., 7(4)*, 529–539.

36. Gupta, P., Kaushik, K., De, S., & Jana, S. (2013). Feasibility analysis on integrated recharging and data collection in pollution sensor networks. In *Proc. Nat. Conf. Commun.*, New Delhi, India.

37. Gupta, V. & De, S. (2018). SBL-based adaptive sensing framework for WSN-assisted IoT applications. *IEEE Internet of Things J., 5(6)*, 4598–4612.

38. Gupta, V., Tripathi, S., & De, S. (2020). Green Sensing and communication: A step towards sustainable IoT systems. *J. Indian Inst. Sc. 100(2)*, 383–398.

39. Kumar, N., Dhand, T., Jindal, A., Aujla, G. S., Cao, H., & Yang, L. (2020). An Edge-Fog Computing Framework for Cloud of Things in Vehicle to Grid Environment. In *2020 IEEE 21st International Symposium on "A World of Wireless, Mobile and Multimedia Networks" (WoWMoM)*, Cork, Ireland, (pp. 354–359).

40. Gupta, V. & De, S. (2020). Collaborative multi-sensing in energy harvesting wireless sensor networks. *IEEE Trans. Signal Inf. Process. Netw., 6(1)*, 426–441.

41. He, Y., Cheng, X., Peng, W., & Stuber, G. L. (2015). A survey of energy harvesting communications: Models and offline optimal policies. *IEEE Commun. Mag., 53(6)*, 79–85.

42. Jelicic, V., Magno, M., Brunelli, D., Paci, G., & Benini, L. (2013). Context-adaptive multimodal wireless sensor network for energy-efficient gas monitoring. *IEEE Sensors J., 13(1)*, 328–338.

43. Kansal, P. & Bose, A. (2012). Bandwidth and latency requirements for smart transmission grid applications. *IEEE Trans. Smart Grid, 3(3)*, 1344–1352.

44. Kaushik, K., Mishra, D., De, S., Basagni, S., Heinzelman, W., Chowdhury, K., & Jana, S. (2013). Experimental demonstration of multi-Hop RF energy transfer. In *Proc. IEEE PIMRC*, London, UK.

45. Kaushik, K., Mishra, D., De, S., Chowdhury, K., & Heinzelman, W. (2016). Low-cost wake-up receiver for RF energy harvesting wireless sensor networks. *IEEE Sensors J., 16(16)*, 6270–6278.

46. Kaushik, K., Mishra, D. & De, S. (2019). Stochastic solar harvesting characterization for sustainable sensor node operation. *IET Wireless Sensor Systems J., 9(4)*, 208–217.

47. Khan, J., Bhuiyan, S., Murphy, G., & Williams, J. (2016). Data denoising and compression for smart grid communication. *IEEE Trans. Signal Inf. Process. Netw., 2(2)*, 200–214.

48. Saleem, A., Khan, A., Malik, S. U. R., Pervaiz, H., Malik, H., Alam, M., & Jindal, A. (2020). FESDA: Fog-Enabled Secure Data Aggregation in Smart Grid IoT Network. *IEEE Internet of Things J., 7(7)*, 6132–6142.

49. Kozłowski, A. & Sosnowski, J. (2019). Energy efficiency tradeoff between duty-cycling and wake-up radio techniques in IoT networks. *Wireless Pers. Commun., 107(4)*, pp. 1951-1971, 1–21.

50. Kumar, S., De, S., & Mishra, D. (2018). RF energy transfer channel models for sustainable IoT. *IEEE Internet of Things J., 5(4)*, 2817–2828.

51. Liu, C., McArthur, S., & Lee, S. (2016). *Smart Grid Handbook, 1*, John Wiley & Sons, chs. 10–11.

52. Lien, S. Y., Hung, S. C., Deng, D. J., & Wang, Y. J. (2017). Efficient ultra-reliable and low latency communications and massive machine-type communications in 5G new radio. In *Proc. IEEE GLOBECOM*, Singapore, 1–7.

53. Lu, H., Brush, A. J. B., Priyantha, B., Karlson, A. K., & Liu, J. (2011). SpeakerSense: Energy efficient unobtrusive speaker identification on mobile phones. In *Proc. Intl. Conf. Pervasive Computing*, Springer-Verlag, Berlin, Heidelberg, 188–205.

54. Marsan, M. A., Bucalo, G., Di Caro, A., Meo, M., & Zhang, Y. (2013). Towards zero grid electricity networking: Powering BSs with renewable energy sources. In *Proc. IEEE ICC*, Budapest, Hungary, 596–601.

55. Mishra, D., Kaushik, K., De, S., Basagni, S., Chowdhury, K., Jana, S., & Heinzelman, W. (2014). Implementation of multi-path energy routing. In *Proc. IEEE PIMRC*, Washington, DC, USA.

56. Mishra, D. & De, S. (2015). Optimal relay placement in two-hop RF energy transfer. *IEEE Trans. Commun., 63(5)*, 1635–1647.

57. Mishra, D., De, S. & Chowdhury, K. (2015). Charging time characterization for wireless RF energy transfer. *IEEE Trans. Circuits Syst.–II: Express Briefs, 62(4)*, 362–366.

58. Morstyn, T., Hredzak, B., Demetriades, G. D., & Agelidis, V. G. (2016). Unified distributed control for DC microgrid operating modes. *IEEE Trans. Power Sys., 31(1)*, 802–812.

59. Mudumbai, R. & Dasgupta, S. (2014). Distributed control for the smart grid: The case of economic dispatch. In *Proc. Inf. Th. App. Wksp. (ITA)*, San Diego, CA, USA, 1–6.

60. Mukherjee, P. & De, S. (2020). A system state aware switched-multichannel protocol for energy harvesting CRNs. *IEEE Trans. Cognitive Commun. Netw., 6(2)*, 669–682.

61. Nabavi, S., Zhang, J., & Chakrabortty, A. (2015). Distributed optimization algorithms for wide-area oscillation monitoring in power systems using interregional PMU-PDC architectures. *IEEE Trans. Smart Grid, 6(5)*, 2529–2538.

62. Nath, S. (2013). ACE: Exploiting correlation for energy-efficient and continuous context sensing. *IEEE Trans. Mobile Comput., 12(8)*, 1472–1486.

63. Oh, E., Son, K., & Krishnamachari, B. (2013). Dynamic base station switching-on/off strategies for green cellular networks. *IEEE Trans. Wireless Commun., 12(5)*, 2126–2136.

64. Oller, J., Demirkol, I., Casademont, J., Paradells, J., Gamm, G., & Reindl, L. (2015). Has time come to switch from duty-cycled mac protocols to wake-up radio for wireless sensor networks? *IEEE/ACM Trans. Netw., 24(2)*, 674–687.

65. Osama, D., Ghoneim, A., & Manjaunath, B. R. (2015). Air pollution clustering using K means algorithm in smart city. *Intl. J. Innovative Res. Comput. Commun. Eng., 3(7)*, 51–57.

66. Patel, M., Aivaliotis, S., Allen, E., et al. (2010). Real-time application of synchrophasors for improving reliability. Princeton, NJ, USA, *Tech. Rep.*

67. Priyantha, B., Lymberopoulos, D., & Liu, J. (2011). LittleRock: Enabling energy-efficient continuous sensing on mobile phones. *IEEE Pervasive Comput., 10(2)*, 12–15.

68. Pudlewski, S., Prasanna, A., & Melodia, T. (2012). Compressed-sensing-enabled video streaming for wireless multimedia sensor networks. *IEEE Trans. Mob. Comput., 11(6)*, 1060–1072.

69. Rao, K. R. & Yip, P. (1990). Discrete Cosine Transform: Algorithms, Advantages, Applications. San Diego, CA, USA: *Academic Press Professional, Inc.*, ch. 7.

70. Richter, F., Fehske, A. J., & Fettweis, G. P. (2009). Energy Efficiency Aspects of Base Station Deployment Strategies for Cellular Networks. In *Proc. IEEE VTC Fall*, Anchorage, AK, USA, 1–5.

71. Pullum, L. L., Jindal, A., Roopaei, M., et al. (2018). Big Data Analytics in the Smart Grid: Big Data Analytics, Machine Learning and Artificial Intelligence in the Smart Grid: Introduction, Benefits, Challenges and Issues. IEEE.

72. Aujla, G.S., Garg, S., Batra, S., Kumar, N., You, I. and Sharma, V., 2019. DROpS: A demand response optimization scheme in SDN-enabled smart energy ecosystem. Information Sciences, 476, pp.453-473.

73. Ringwelski, M., Renner, C., Reinhardt, A., Weigel, A., & Turau, V. (2012). The hitchhiker's guide to choosing the compression algorithm for your smart meter data. In *Proc. IEEE Intl. Energy Conf. Exhibit.*, 935–940.

74. Roy Chowdhury, M., Shukla, N. K., De, S., & Biswas, R. (2018). Energy-efficient air pollution monitoring with optimum duty-cycling on a sensor hub. In *Proc. Nat. Conf. Commun.*, Hyderabad, India.

75. Roy Chowdhury, M., Tripathi, S., & De, S. (2020). Adaptive multivariate data compression in smart metering Internet of Things. *IEEE Trans. Ind. Informat.*

76. Salem, Q., Liu L., & Xie, J. (2018). Islanding and Resynchronization Process of a Grid-Connected Microgrid with Series Transformerless H-Bridge Inverter Installed at PCC. In *Proc. IEEE EEEIC and I&CPS Europe*, Palermo, Italy, 1–6.

77. Salomon, D. (2007). Data Compression: The Complete Reference. *Springer London*, ch. 4.

78. Sangare, F. et al. (2017). Mobile charging in wireless-powered sensor networks: Optimal scheduling and experimental implementation. *IEEE Trans. Veh. Technol., 66(8)*, 7400–7410.

79. Sezer, O. B., Dogdu, E., & Ozbayoglu, A. M. (2017). Context aware computing, learning and big data in Internet of Things: A survey. *IEEE Internet of Things J.*

80. Sheng, X., Tang, J., & Zhang, W. Energy-efficient collaborative sensing with mobile phones. In *Proc. IEEE INFOCOM*, 1916–-1924.

81. Singhal, C., Kumar, S., De, S., Panwar, N., Tonde, R., & De, P. (2014). Class-based shared resource allocation for cell-edge users in OFDMA networks. *IEEE Trans. Mob. Comput., 13(1)*, 48–60.

82. Singhal, C., De, S., Trestian, R. & Muntean, G.-M. (2014). Joint optimization of user-experience and energy-efficiency in wireless multimedia broadcast. *IEEE Trans. Mob. Comput., 13(7)*, 1522–1535.

83. Singhal, C., De, S., Trestian, R., & Muntean, G.-M. (2015). eWU-TV: User-centric energy-efficient digital TV broadcast over Wi-Fi networks. *IEEE Trans. Broadcast., 61(1)*, 39–55.

84. Singhal, C. & De, S. (2016). Energy-efficient and QoE-aware TV broadcast in next-generation heterogeneous networks. *IEEE Commun. Mag., 54(12)*, 142–150.

85. Singhal, C. & De, S. (2018). eM-SON: Efficient multimedia service over self-organizing Wi-Fi network. *IET Netw. J., 7(4)*, 181–189.

86. Singhal, C. & De, S. (2019). UE-TV: User-centric energy-efficient HDTV broadcast over LTE and Wi-Fi. *IEEE Trans. Mobile Comput., 18(7)*, 1703–1717.

87. Spenza, D., Magno, M., Basagni, S., Benini, L., Paoli, M., & Petrioli, C. (2015). Beyond duty cycling: Wakeup radio with selective awakenings for long-lived wireless sensing systems. In *Proc. IEEE INFOCOM*, Hong Kong, China, 522–530.

88. Jindal, A., Kumar, N., & Singh, M. (2020). A unified framework for big data acquisition, storage, and analytics for demand response management in smart cities. Future Generation Computer Systems, 108, 921-934.

89. Sotomayor-Olmedo, A., Aceves-Fernndez, M. A., Gorrostieta-Hurtado, E., Pedraza-Ortega, C., Ramos-Arregun, J. M., & Vargas-Soto, J. E. (2013). Forecast urban air pollution in Mexico city by using support vector machines: A kernel performance approach. *Intl. J. Intelligence Science, 3*, 126–135.

90. Stephen, B., Mutanen, A. J., Galloway, S., Burt, G., & Jarventausta, P. (2014). Enhanced load profiling for residential network customers. *IEEE Trans. Power Del., 29(1)*, 88–96.

91. Su, W., Wang, J., & Roh, J. (2014). Stochastic energy scheduling in microgrids with intermittent renewable energy resources. *IEEE Trans. Smart Grid, 5(4)*, 1876–1883.

92. Suman, S. & De, S. (2017). Solar-enabled green base stations: Cost versus utility. In *Proc. IEEE WoWMoM*, Macau, China.

93. Suman, S., Kumar, S., & De, S. (2018). Path loss model for UAV-assisted RFET. *IEEE Commun. Lett., 22(10)*, 2048–2051.

94. Aujla, G.S.S., Kumar, N., Garg, S., Kaur, K. and Ranjan, R., 2019. EDCSuS: sustainable edge data centers as a service in SDN-enabled vehicular environment. *IEEE Trans. Sustain. Comput.* DOI: 10.1109/TSUSC.2019.2907110

95. Aujla, G.S., Singh, M., Kumar, N. and Zomaya, A., 2017. Stackelberg game for energy-aware resource allocation to sustain data centers using RES. *IEEE Transactions on Cloud Comput., 7(4)*, 1109–1123.

96. Aujla, G.S. and Kumar, N., 2018. MEnSuS: An efficient scheme for energy management with sustainability of cloud data centers in edge–cloud environment. *Future Generation Computer Systems, 86*, 1279–1300.

97. Suman, S. & De, S. (2019). Low complexity dimensioning of sustainable solar-enabled systems: A case of base station. *IEEE Trans. Sustain. Comput., 5(3)*, 438-454.

98. Suman, S., Kumar, S., & De, S. (2019). UAV-assisted RFET: A novel framework for sustainable WSN. *IEEE Trans. Green Commun. Netw., 3(4)*, 1117–1131.

99. S., Kumar, S., & De, S. (2019). Impact of hovering inaccuracy on UAV-aided RFET. *IEEE Commun. Lett., 23(12)*, 2362–2366.

100. Tai, A. P. K., Mickley, L. J., & Jacob, D. J. (2010). Correlations between fine particulate matter (PM2.5) and meteorological variables in the United States: Implications for the sensitivity of PM2.5 to climate change. *Atmospheric Environment, 44(32)*, 3976–3984.

101. Tate, J. E. (2016). Preprocessing and golomb -rice encoding for lossless compression of phasor angle data. *IEEE Trans. Smart Grid, 7(2)*, 718–729.

102. Tcheou, M. P., et al. (2014). The compression of electric signal waveforms for smart grids: State of the art and future trends. *IEEE Trans. Smart Grid, 5(1)*, 291—302.

103. Tong, X., Kang, C., & Xia, Q. (2016). Smart metering load data compression based on load feature identification. *IEEE Trans. Smart Grid, 7(5)*, 2414–2422.

104. Tripathi, S. & De, S. (2018). Dynamic prediction of powerline frequency for wide area monitoring and control. *IEEE Trans. Ind. Informat., 14(7)*, 2837–2846.

105. Tripathi, S. & De, S. (2018). An efficient data characterization and reduction scheme for smart metering infrastructure. *IEEE Trans. Ind. Informat., 14(10)*, 4300–4308.

106. Tripathi, S. & De, S. (2019). Data-driven optimizations in IoT: A new frontier of challenges and opportunities. *Springer CSI Trans. ICT, 7(1)*, 35–43.

107. Tripathi, S. & De, S. (2020). Channel-adaptive transmission protocols for smart grid IoT communication. *IEEE IoT J.*

108. Tiwari, S., Bisht, D. S., Srivastava, A. K., Pipal, A. S., Taneja, A., Srivastava, M. K., & Attri, S. D. (2014). "Variability in atmospheric particulates and meteorological effects on their mass concentrations over Delhi, India. *Atmospheric Research, 145–146*, 45–56.

109. Unterweger, A. & Engel, D. (2015). Resumable load data compression in smart grids. *IEEE Trans. Smart Grid, 6(2)*, 919–929.

110. Wang, Y., Lin, J., Annavaram, M., Jacobson, Q. A., Hong, J., Krishnamachari, B., & Sadeh, N. (2009). A framework of energy efficient mobile sensing for automatic user state recognition. In *Proc. ACM MobiSys*, New York, NY, USA, 179–192.

111. Wang, C., Li, J., Ye, F., & Yang, Y. (2016). A mobile data gathering framework for wireless rechargeable sensor networks with vehicle movement costs and capacity constraints. *IEEE Trans. Comput., 65(8)*, 2411–2427.

112. Wang, Y., Chen, Q., Kang, C., Xia, Q., & Luo, M. (2017). Sparse and redundant representation-based smart meter data compression and pattern extraction. *IEEE Trans. Power Syst., 32(3)*, 2142–2151.

113. Wen, L., Zhou, K., Yang, S., & Li, L. (2018). Compression of smart meter big data: A survey. *Renew. Sustain. Energy Rev., 91*, 59—69.

114. Xie, L., Shi, Y., Hou, Y. T., & Sherali, H. D. (2012). Making sensor networks immortal: An energy-renewal approach with wireless power transfer. *IEEE/ACM Trans. Netw., 20(6)*, 1748–1761.

115. Xie, L., Chen, Y., & Kumar, P. R. (2014). Dimensionality reduction of synchrophasor data for early event detection: Linearized analysis. *IEEE Trans. Power Syst., 29(6)*, 2784–2794.

116. Zeinali, M. & Thompson, J. S. (2016). Impact of compression and aggregation in wireless networks on smart meter data. In *Proc. IEEE Intl. Wksp. Sig. Process. Adv. Wireless Commun.*, 1–5.

117. Jindal, A., Kronawitter, J., Kuhn, R., Bor, M., de Meer, H., Gouglidis, A., Hutchison, D., Marnerides, A. K., Scott, A., & Mauthe, A. (2020). A flexible ICT architecture to support ancillary services in future electricity distribution networks: an accounting use case for DSOs. Energy Informatics, 3(1), 1-10.

118. Zeng, Y., et al. (2016). Wireless communications with unmanned aerial vehicles: opportunities and challenges. *IEEE Commun. Mag., 54(5)*, 36–42.

119. Zeng, S. & Zhang, Y. (2017). The effect of meteorological elements on continuing heavy air pollution: A case study in the Chengdu area during the 2014 Spring Festival. *Atmosphere, 8(4)*, 1–17.

120. Zhang, Y., Gatsis, N., & Giannakis, G. B. (2013). Robust energy management for microgrids with high-penetration renewables. *IEEE Trans. Sust. Energy, 4(4)*, 944–953.

121. Zhao, F., Shin, J., & Reich, J. (2002). Information-driven dynamic sensor collaboration. *IEEE Signal Proc. Mag., 19(2)*, 61–72.

122. Zhou, Z., Chawla, N. V., Jin, Y., & Williams, G. J. (2014). Big data opportunities and challenges: Discussions from data analytics perspectives. *IEEE Comput. Intell. Mag., 9(4)*, 62–74.